INTO THE GROOVE

Some other titles in the Bloomsbury Sigma series:

Sex on Earth by Jules Howard
Spirals in Time by Helen Scales
A is for Arsenic by Kathryn Harkup
Suspicious Minds by Rob Brotherton
Herding Hemingway's Cats by Kat Arney
The Tyrannosaur Chronicles by David Hone
Soccermatics by David Sumpter
Wonders Beyond Numbers by Johnny Ball
The Planet Factory by Elizabeth Tasker
Seeds of Science by Mark Lynas
Turned On by Kate Devlin
We Need to Talk About Love by Laura Mucha
Borrowed Time by Sue Armstrong
The Vinyl Frontier by Jonathan Scott
Clearing the Air by Tim Smedley
The Contact Paradox by Keith Cooper
Life Changing by Helen Pilcher
Sway by Pragya Agarwal
Kindred by Rebecca Wragg Sykes
Our Only Home by His Holiness The Dalai Lama
First Light by Emma Chapman
Models of the Mind by Grace Lindsay
The Brilliant Abyss by Helen Scales
Overloaded by Ginny Smith
Beasts Before Us by Elsa Panciroli
Our Biggest Experiment by Alice Bell
Aesop's Animals by Jo Wimpenny
Fire and Ice by Natalie Starkey
Sticky by Laurie Winkless
Racing Green by Kit Chapman
Wonderdog by Jules Howard
Growing Up Human by Brenna Hassett
Superspy Science by Kathryn Harkup

INTO THE GROOVE

The Story of Sound From Tin Foil to Vinyl

Jonathan Scott

BLOOMSBURY SIGMA
LONDON · OXFORD · NEW YORK · NEW DELHI · SYDNEY

BLOOMSBURY SIGMA
Bloomsbury Publishing Plc
50 Bedford Square, London, WC1B 3DP, UK
29 Earlsfort Terrace, Dublin 2, Ireland

BLOOMSBURY, BLOOMSBURY SIGMA and the Bloomsbury Sigma logo are trademarks of Bloomsbury Publishing Plc

First published in the United Kingdom in 2023

Copyright © Jonathan Scott, 2023

Jonathan Scott has asserted his rights under the Copyright, Designs and Patents Act, 1988, to be identified as Author of this work

All rights reserved. No part of this publication may be reproduced or transmitted in any form or by any means, electronic or mechanical, including photocopying, recording, or any information storage or retrieval system, without prior permission in writing from the publishers

Bloomsbury Publishing Plc does not have any control over, or responsibility for, any third-party websites referred to or in this book. All internet addresses given in this book were correct at the time of going to press. The author and publisher regret any inconvenience caused if addresses have changed or sites have ceased to exist, but can accept no responsibility for any such changes

A catalogue record for this book is available from the British Library

Library of Congress Cataloguing-in-Publication data has been applied for

ISBN: HB: 978-1-4729-7982-7; eBook: 978-1-4729-7980-3

2 4 6 8 10 9 7 5 3 1

Typeset by Deanta Global Publishing Services, Chennai, India
Chapter illustration © CSA Images/Getty
Printed and bound in Great Britain by CPI Group (UK) Ltd, Croydon CR0 4YY

Bloomsbury Sigma, Book Seventy-Seven

To find out more about our authors and books visit www.bloomsbury.com and sign up for our newsletters

Contents

Foreword	8
One: The End	13
Two: The Beginning	25
Three: Oddballs	34
Four: Telegraphs and Telephones	41
Five: The Barb of a Feather	49
Six: The First Phonograph	55
Seven: Fade Away	66
Eight: The Volta Lab	77
Nine: Edison Returns	89
Ten: Performers and Producers	98
Eleven: Crossing Continents	112
Twelve: Berliner, Johnson, Seaman and Jones	126
Thirteen: Coin Slots and Record Shops	137
Fourteen: Tubes and Tone Tests	147
Fifteen: Studios, Scouts and Engineers	158
Sixteen: The Electrical Era	177
Seventeen: The Coming of the 33	189

Eighteen: The Long Players	205
Nineteen: The Speed Wars	218
Twenty: Shows, Soundtracks and Sleeves	226
Twenty-one: Stereo and Magnetic Tape	235
Twenty-two: Discos and Curios	245
Twenty-three: Revolutions in Space	254
Miscellany of the Groove	266
Bibliography	305
Acknowledgements	311
Index	314

For Genorence and Runessa

FOREWORD

'Certainly, within a dozen years, some of the great singers will be induced to sing into the ear of the phonograph, and the stereotyped cylinders thence obtained will be put into the hand organs of the streets, and we shall hear the actual voice of Christine Nilsson or Annie Louise Cary ground out at every corner. In public exhibitions, also, we shall have reproductions of the sounds of nature, and of noises familiar and unfamiliar. Nothing will be easier than to catch the sounds of the waves on the beach, the roar of Niagara, the discords of the street, the voice of animals, the puffing and rush of the railroad, the rolling thunder, or even the tumult of a battle.'

George Prescott, 1879

Thomas Edison was showing a visitor around his laboratories at Menlo Park, New Jersey. The main laboratory building was a long, two-storey rectangular block. There was a glass house, carpenters' shop, carbon shed and a blacksmiths. Tours could be detailed, with lengthy pauses at various inventions, explaining how they worked, what they did, how he had thought of them, and how he and his crack team of engineers had made them. The stars of the show were usually the light bulb, the telephone and the phonograph.

To a layman in the late 1870s, the first sight of the phonograph might well have been an anticlimax. It looked so ordinary. It looked like a lathe. It looked like the kind of thing you'd see in any workshop up and down the land. A machine that could talk surely needed more moving parts, more bells and whistles. Edison took his time, explaining every aspect of his devices, how they capture, relayed or replayed sound. The listener replied with lots of 'oh rights' and 'I sees' and took copious notes. Edison was pleased. The inventor was known as the kind of man who did not suffer fools gladly but would warmly engage with anyone he felt 'got it'. He loved the camaraderie of the workshop, the company of engineers, mechanists and technicians, but would give time to anyone who seemed to understand what he was on about. His explanation came to an end. The man only had one question.

'I understand it all except how the sound gets out again.'

Edison's face fell.

The reason I love records, the thing that inspired me to write this book, is that I don't understand them either. I remain as delighted and mystified by them as when I was five years old. The record player in the house where

I grew up was raised high off the ground in the corner of the living room. We couldn't be trusted to touch the records, even less to put them safely on the turntable and drop the needle. We had a powerful 30W amp that took a while to warm up and then crackled alarmingly whenever you touched the volume. This part of the operation, I was allowed to oversee. When it came to the actual vinyl, I'd have to badger a parent, and they always seemed to take ages to arrive. Perhaps that's what added to the feeling of excitement: the long build-up before a wonderful rush of warm crackling, as the needle skipped once, twice, then found its place in the groove, rich sounds suddenly filling the room. I still remember the feeling of wonder.

There are easier ways to listen to music. I can type the title of a record into AppleMusic or Spotify that took me years to track down in the mid-1990s. A few taps later, I'm listening to it again, and I've barely moved. There's nothing wrong with that, of course, but it's just ... different. With records, the process is slow and delightful – the switches, buttons and knobs, the volume and tone controls, the little red light, the soft hum, the selection of the record, the removal from the sleeve, the careful placing on the turntable, speed select, click, spin ... Vinyl is the format for when you have the time and will to engage with music in a deeper way. It's the format for when you're ready to immerse yourself, to kneel at the altar and just listen. It's not for when you're doing the dishes, or rushing to the shops, or stripping wallpaper, or writing a book. Vinyl is for you. It's the moment that you give yourself, a trip to the cathedral of sound.

I still put on records. I still place the needle on the groove. And I still don't quite get it. How? How does it work? How is it that plastic, albeit plastic arranged in a very specific way, can sing? Not only that, it can sing for ages, and not only that but sing for ages with a full band, and not only that but sound so good it's like the band is right there with you. And don't even get me started on orchestras. Just, how? It's bumpy plastic being rubbed by a pliable spike ... how can that be music? How can one little undulation here, one little rivulet there, be a drum, the other a bass, or a tuba?

Back when I was five, of course, I wouldn't marvel at drum, bass or tuba, but that it sounded *exactly* like Penelope Keith, as it was she who was narrating Hans Christian Andersen's fairy tales. But still, how was it Penelope Keith? I remember asking my father how it worked and afterwards being none the wiser. I must have asked some kind of 'kid question' that prompted a weird or hard-to-understand response, however, as from that day forwards, whenever I thought of records, a particular vision would come to me, of a giant stylus in the sky, being placed by a giant hand, into a giant groove of the Grand Canyon. And it always got me wondering ... what would the Grand Canyon play? Would that sound like Penelope Keith?

It's this feeling of wonder that I want you to hold on to as we go through this story. Because it's a wonder that never leaves me. Even though I 'understand' how it all works now, there's still the same five-year-old me who really does not.

ONE

The End

> 'On looking at the plate more carefully, I noticed a long row of fine streaks parallel and equidistant from one another.'
>
> Galileo Galilei, *Dialogue Concerning the Two Chief World Systems*, 1632

Two weeks before he died, Dr Peter Goldmark was shaking hands with the US president. The Hungarian-born engineer was in Washington, DC, for an awards ceremony, one of several scientists rubbing shoulders as they awaited a speech and a backslap from the leader of the free world. It was mid-morning, 22 November 1977, when President Jimmy Carter entered Room 450 of the Eisenhower Executive Office Building. Each scientist's name was read aloud, followed by a short explanation of the work that had got them there. One had conducted studies into the social organisation of insects; another was being recognised for pioneering metallurgy; yet another for sub-cellular mechanisms. At the mention of Dr Goldmark, the president went off script, interjecting that he was particularly grateful for *his* invention: the beloved, most perfectly pleasing and dearest of things, the long-playing record.

This book is all about how these crackling, fragile beings came into existence. The first records, created almost exactly a hundred years earlier in December 1877, were tube-shaped. The phonographs that played them were dubbed 'talking machines' by the popular press, as that's what they seemed to be: machines that talked. The cylindrical records were made of tin foil, then came wax, before proto-discs appeared, spinning at various speeds, in various sizes and made of various substances, from hardened rubber, to glass, to enamelled paper, until 10- and 12-inch shellac 78rpm records became the dominant format for a generation, eventually giving way to vinyl from the 1940s.

The phonograph was a world-changing device, one that ultimately shifted the way we humans interacted with music. At first, having presented his new invention at the offices of *Scientific American* in New York City in late 1877, Thomas Edison thought it would be of most use in the world of business – as a Dictaphone-type thingy that might do away with the humble stenographer. He wrote about its potential uses, predicting it would become a popular way of recording the voices of loved ones before they shuffled off, and as a means for producing talking books for the blind. The reproduction of music, while on his radar, came fourth on a list of ten.

During the 1890s, the talking machine business took off, nickel slot-machines with listening tubes making good money across America, and electric, hand-cranked and wind-up players becoming increasingly common in middle- and upper-class homes. There were even machines so low cost, such as the German-made, lyre-shaped 'Puck', that virtually any household could enjoy brass bands, whistled ditties and comedic skits on demand.

Music was being reshaped by the phonograph. Before talking machines, virtually all music was either live or learnt – performed in public or practised at home. The phonograph created a new intimacy between music and listener. Punters could now consume music at home, alone if they wished. Indeed, they could listen and listen again, they could study and debate, love and obsess, become collectors and musos, affix themselves to types of music, become fans, enjoy the chills, shivers and goosebumps set off by refrain, phrase, voice, harmony, key and chord.

However, not everyone welcomed the invention. Musicians grumbled that the upstart phonograph would gut live performance and didn't pay them royalties. The first recording star, American 'march king' John Philip Sousa, would later write a letter bemoaning the menace of canned music, and how music on demand would rob children of the desire to learn instruments. Many artists didn't much care for the quality of reproduction, and composers hated the short format. The capacity of cylinders, and the first generations of discs that followed in their wake, was very limited indeed – between two and four minutes per side. This only allowed for the most heavily abridged versions of symphonies, concertos, tone poems and chamber music, but it did help popularise dance music and the three-minute song.

The acoustic era of the fledgling recording industry lasted until 1925. After that date, performances captured by microphone could be amplified electrically before being cut into the grooves of a record. Prior to that date, all recording was acoustic, and frankly, not everything sounded good when recorded through a horn. Many early singers so disliked how they sounded on wax that they refused to perform, and those who did often stumbled amid the unfamiliar pressure of getting it right first time on demand – in the Victorian recording studio a clean take was paramount. Others felt humiliated that, to put any emotion into a performance, they had to stick their heads practically inside the recording horn for the quiet bits, before leaping back for the high Cs. There were musicians whose instruments simply didn't translate well in the limited frequency range of the acoustic era, and so couldn't make any money from the new racket. And in that first iteration of the recording industry, there was no way of duplicating a record, so to sell another recording of a particular performance, the performer had to perform the damn thing all over again. Indeed, the only way to 'mass produce' was to have more than one recording device set up. So brass bands, which made many of the radio-friendly unit shifters of the late 1800s, might perform 'The Liberty Bell' before a carefully arranged rank of 10 recording horns, each leading to a diaphragm and carving stylus, producing 10 cylinders in one go. Of course, any mistakes meant 10 wasted cylinders in one go, which then had to be shaved and remounted before the whole laborious process started again.

Naming the inventor of a specific thing can be a tricky business. Often credit goes not to the originator of an idea, or even the creator of the first machine, but to the inventor who builds a working prototype, the more reliable and cheaper to produce the better. Then there's the problem of defining which idea, tweak or prototype is the most valuable in the series of evolutionary steps that culminate in an object. An idea without development can end up being useless. The telephone gives us a fine example. Alexander Graham Bell's most ardent fan would admit that, while he rightly claims to be its inventor, there were others working in the same arena who managed to create very similar working tech around the same time. But as they weren't as well placed or well backed, history left them behind.

But the phonograph is different. Thomas Edison is known as the inventor of the phonograph … and he was. While working on improvements to the telegraph and telephone, he came up with an idea, tested it, knocked up a working prototype with the help of his crack team of engineers at Menlo Park, New Jersey, and launched the first generation of sound-reproducing machines using tin foil, which did what they were supposed to do after a fashion.

Perhaps because he has so many other credits to his globally famous name – and so it feels a bit tedious to say 'you know that guy who you already knew invented the phonograph? Well he did!' – sound historians are fond of mentioning his name with caveats or stressing the pioneering work of contemporaries and those who came before. Happily, for us, comfortably strapped into our reclining seats and about to embark on this weird journey into sound, there are plenty of them to choose from.

For starters, Thomas Edison was the first to *play back* sound, but he was not the first to *record* sound, not by a long way. Twenty years before the phonograph, there was the phonautograph, a device patented in 1857 by Édouard-Léon Scott de Martinville, which at first glance looks a bit like an exaggerated steampunk-style phonograph or perhaps some kind of device for de-tangling wool. The phonautograph made phonautograms, strips of paper on which sound was recorded as a traced squiggle.

Édouard-Léon was a printer and typesetter by trade. If you consider a typesetter of the day had to be adept at spotting errors in text upside down and back to front, it is unsurprising to learn he lived in hope that his phonautograms could become the daguerreotypes of sound, the photography of speech. He hoped that with practice they could be 'read', that they might one day partially replace the printed word or at the very least become a new form of stenography. While he was wrong in this respect, he would in a sense be proved right 150 years later, when a team at the Library of Congress managed to reconstruct some of his phonautograms, successfully bringing his own unearthly singing of 'Au clair de la lune' in April 1860 back to life. In other words, they resurrected humanity's earliest recorded voice, a generation before Edison.

You'll often hear the name Charles Cros mentioned in the tale of vinyl. That's because just before Edison recorded *his* idea in a laboratory notebook, this marvellously romantic, absinthe-swigging Parisian poet came up with a similar, scientifically sound machine he called the paleophone, which could also record and reproduce sound. However, being too strapped for cash or contacts to make the idea reality, it existed only on paper, in a sealed letter that he sent to the Académie des Sciences in Paris in April 1877.

Edison was not the first to perfect or improve his machine, either. The first-generation phonographs were wonders of the age, yes, but they were hard to use, unreliable and sounded sketchy as all hell. By the end of the 1870s Edison had turned his attention elsewhere, to more profitable projects, leaving a decade-long hiatus during which the world was a place where humans could record and reproduce sounds but hardly any of them ever did.

Into this void emerged Alexander Graham Bell. Partly born of frustration that he wasn't the first to come up with the phonograph, and partly because his father-in-law wasn't getting any return from his investment in Edison's tin-foil business, Bell opened Victorian hi-fi's most fascinating institution – the Volta Laboratory. In this hub of research and development, located in Washington, DC, a band of engineers tried out almost every method imaginable for carving sound onto a surface. They tried numerous world-firsts in the field of acoustics, testing out all sorts of bizarre set-ups with metal, magnetism, water, light and glass, and – as we know from the reconstruction of fragmentary discs and cylinders that survive – committing to disc what is possibly humanity's oldest surviving swear word.

Back to Dr Peter Goldmark, who was working at CBS in the 1940s when he stepped up to the decades-long problem of capacity. His work would kick-start the modern era of vinyl, with ingenious microgroove technology that increased the amount of music that could be squeezed onto a single 12-inch disc by a factor of six. At the same time the new, lighter and more flexible Vinylite plastic greatly reduced the surface noise that had plagued shellac.

Just days after Goldmark was honoured in the US capital, Carl Sagan also paid a visit to President Carter. At an informal White House soirée, they discussed their most recent collaboration – the Voyager Golden Record, a 16rpm LP, a copy of which NASA had just fixed to the Voyager 1 and Voyager 2 probes and launched into space. These probes were embarking on their Grand Tour of the solar system, after which they were destined to drift forever in the unimaginable void of interstellar space, speeding away from our sun at around 38,000mph. The records were conceived as a message, a kind of multimedia guide to Earth and its Earthlings, should the probes ever be found by intelligent beings. And they included these words from President Carter:

> This is a present from a small distant world, a token of our sounds, our science, our images, our music, our thoughts, and our feelings. We are attempting to survive our time so we may live into yours. We hope someday, having solved the problems we face, to join a community of galactic civilizations. This record represents our hope and our determination, and our good will in a vast and awesome universe.

What I'm stressing here is that 1977 was a pivotal year for vinyl. Just a hundred years on from Edison's tin foil, and the format had gone interstellar. The Golden Records were a pinnacle of sorts – the groove pushed to its limit, stretching the capacity to include 90 precious minutes of breathtaking music, alongside poetry, whale song, greetings, sounds and even photographs, all cast into the cosmos aboard a golden arc that will in all likelihood outlast our planet. It was also the year vinyl sales peaked in the United States, the year of *Rumours*, *Bat Out of Hell* and *Never Mind the Bollocks, Here's the Sex Pistols*. Some of the best-sounding sides ever cut were put out on reassuringly heavy, thick, hard-to-warp discs that sounded rich and warm on your turntable. Yet even then these were giving way to cheaper, thinner, more flimsy records, with tinny, scratchy sounds to match. As well as the death of Dr Goldmark, 1977 saw the passing of Elvis Presley, whose early career straddled the format wars and the arrival of the 45, and Bing Crosby, the first global star to benefit from technology that could capture his crooning voice and make it sound good on record.

But we're beginning to venture dangerously close to the run-out groove of this story. For now, let's lift the tonearm, and place the needle back to the beginning of side A.

The story of patent number 372,786 really gets going with Galileo, who was running an iron chisel over a brass plate to remove imperfections sometime before the year 1632. The first thing he noticed was a strong, clear whistling sound, the second was the dust, which began to dance across the brass and gather in regular patterns. 'On looking at the plate more carefully, I noticed a long row of fine streaks parallel and equidistant from one another,' he wrote. Fifty years later Robert Hooke covered some glass plates with flour, ran a violin bow along the edge of the glass, and watched as more distinctive patterns emerged. Then a trumpet player named John Shore came up with the tuning fork. This ingenious little device could emit a pure tone at an exact frequency when its ends were struck, and simply pressing the handle to a resonating surface amplified the tone. This modest, pocket-sized piece of metal could be dropped, drenched, scratched, thrown pell-mell into any bag or case, and yet still sing in perfect pitch on demand, making it a world-changing device for performers.

Galileo's dust, Hooke's flour and Shore's fork sit on a line that charts our understanding of acoustics and sound. Thinkers and tinkerers soon began experimenting with resonating instruments. The verrillon would sing notes created by filling glasses with water and striking the glasses with wooden hammers. A public performance so delighted Benjamin Franklin on a visit to London in 1757 that he would go on to invent his own version, the glass armonica.

Then we come to Ernst Chladni, who would become known as the father of acoustics. The father of the father of acoustics didn't like his son's interest in science. Ernst junior was from a line of academics – his great-grandfather was a Lutheran clergyman, his grandfather a professor of theology, his uncles were professors, and his father was law professor and rector at the University of Wittenberg. At first he complied, abandoning any interest in science and studying law at the University of Leipzig. But following his father's death, he established himself as a lecturer at Wittenberg and began to pursue his passion in earnest, conducting a series of experiments that ultimately led to the publication of *Discoveries in the Theory of Sound* in 1787.

Chladni mounted iron plates on a single sprocket and covered the surface with fine sand. He would then bow the plate. As it resonated, the sand would divide into discernible regions, vibrations sending the grains scurrying off in opposite directions like children picking teams, settling along nodal lines – areas of 'zero displacement'. Theoretically, any plate has many possible vibration modes, each corresponding to a specific frequency of sound. Each mode produces a unique pattern, the complexity of which increases with the frequency of the vibration. The shape of the patterns produced on a given plate depends on other factors too, including the shape of the plate itself.

These distinctive and repeatable patterns – a cross, a square, a star, a star with curved edges, a perfect circle, a series of wavy lines – were shown to the world in a series of diagrams folded into *Theory of Sound* and became known as Chladni figures. Chladni's live demonstrations became a modest sensation. One took place at the Paris Academy in 1808, where the audience included Napoleon, who would go on to finance a French edition of *Die Akustic*. Napoleon also set a prize for the best mathematical explanation of this mysterious phenomenon. French mathematician Sophie Germain's answer, although rejected due to flaws, was the only entry that took the correct approach.

The success of Chladni's research, combined with the popularity of his public demonstrations, inspired many other acoustic researchers to build on his work. Another common nineteenth-century experiment was to co-sound – to show how one struck tuning fork placed close to another would cause the second one to join in. In 1807 polymath Thomas Young unveiled his 'vibrograph', a device used to measure the frequency of a tuning fork. This was a soot-covered cylinder, mounted on a spindle, set to revolve with weights on a string. The tuning fork would carve a line into the soot as the cylinder slowly descended. French mathematician Jean-Marie Constant Duhamel refined the machine, coming up with the 'vibroscope', notable for the fact that it recorded the vibrations laterally, using a feedscrew – which would be a feature of the first generation of phonographs.

Chladni's patterns captured the imagination of the science community, but also offered the unscientific layperson a vision of sound. These continuing experiments were proving that sound waves travelled through the air, and that if they were strong enough, they left a trace. If they left a trace, this offered the tantalising possibility that a sound could perhaps, one day, be captured, recorded, read or even reconstructed.

Meanwhile, in post-revolutionary France ...

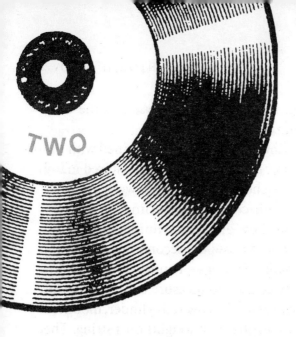

TWO

The Beginning

> '*Gentlemen, we are in the presence of an invention being born …*'
>
> Édouard-Léon Scott de Martinville

If you had walked the Rue Taranne in Paris in the spring of 1860, you might have heard strange noises coming from one of the houses – unnaturally slow, mournful singing; tuning forks sounding; slow and deliberate enunciated poetry. If, overcome with curiosity, you had put your ear to the door, you might have heard another sound – the sound of something mechanical.

You're eavesdropping at the home of Édouard-Léon Scott de Martinville. A typesetter and bookseller by trade, by 1860 he had become completely obsessed with an idea that he thought of as a new 'photography of sound', which he hoped would have the same seismic impact in sound that daguerreotypes were having in light.

The nineteenth century had seen an explosion in printed matter, fuelled by quicker steam-powered presses and rising literacy rates. Whether churning out pamphlets, multi-volume novels, magazines or newspapers, typesetting remained the bottleneck in the print production process. This was a demanding occupation. It required a high level of skill, both in spelling and grammar but also manual dexterity. A good typesetter tended to be well educated and practical, spending hours at a wooden tray filled with cast metal letters and symbols. The compositor would assemble them into words, lines, paragraphs and pages, which would then be tightly bound together in a frame, inked and pressed to paper. The best typesetters and compositors were therefore not only physically competent but able to quickly read text back to front and upside down. With all this in mind, it is logical that a professional typesetter would be the one to come up with a labour-saving innovation that he believed might render the entire time-consuming, fiddly business obsolete.

The idea first came to Scott de Martinville around 1853. This was a period of change within the French capital. Napoleon III's prefect of the Seine, Georges-Eugène Haussmann, had begun a massive public works project, building new boulevards and parks, theatres, markets and monuments across the city. Our typesetter was working at a scientific publishing house, specifically on a volume on human physiology, which included diagrams of the human ear.

He asked himself, if photography was able to capture light using an instrument that copied the human eye, what was to stop him using the human ear as a template for an instrument to capture sound? This wasn't a unique thought. A Parisian theatre critic predicted in a piece written in 1847 that humans would soon be preserving 'sonic undulations' and hanging them on walls, and many others wrote along similar lines, imagining a new photography of sound to be just around the corner. But Scott de Martinville was one of the first to act on the idea. And in the wake of the invention of the phonograph a few years later, he would write a book about his work, perhaps in a bid to ensure his contribution was not forgotten.

The earliest account of his experiments survive in a letter he wrote to the Académie des Sciences in 1857. In the letter he describes covering a plate of glass with a thin layer of lampblack (a finely powdered black soot used chiefly in pigments) and fixing an acoustic trumpet with an eardrum-inspired membrane the diameter of a five-franc coin at its small end. To the centre of this membrane he affixed a stylus – in this case, a boar's bristle, around a centimetre in length.

> I carefully adjust the trumpet so the stylus barely grazes the lampblack. Then, as the glass plate slides horizontally in a well-formed groove at a speed of one metre per second, one speaks in the vicinity of the trumpet's opening, causing the membranes to vibrate and the stylus to trace figures.

This first device had a basic flat-bed design, where the glass moved horizontally, and so could record just a microsecond of sound. With some financial support from the Société d'encouragement pour l'industrie nationale, Scott de Martinville began experimenting with improved prototypes, using parchment membranes, with the boar's bristle now scratching out sound waves on smoked paper wrapped around a rotating cylinder. This design allowed much longer passages of sound to be recorded and, once unwrapped, resulted in long thin undulating lines. He called his machine the phonautograph and the notations 'phonautograms'.

By 1860 Scott had made a number of recordings. He sang songs, he recited excerpts of poetry and plays in several languages. He was convinced this natural calligraphy could be 'read', if only we trained ourselves to read it. And so he put energy into beginning to interpret the sounds, transcribing words spoken beneath the corresponding shapes traced in the paper in an attempt to recognise patterns.

'Gentlemen, we are in the presence of an invention being born,' he wrote, 'an entirely new graphic art springing from the heart of physics, of physiology, of mechanics. Each trace of speech that I submit today analyses the voice: its tonality, its intensity, its timbre. I believe a synthesis is also possible through which the tracing of the words is transformed into a series of signs by mechanical means, and I propose to attempt it. I see the book of nature opened before the gaze of all men, and, however small I may be, I dare hope to be permitted to read it.'

THE BEGINNING

With these first (unplayable) records came the world's first (unplayable) record collector, a largely forgotten Philadelphia merchant named Charles N. Bancker. Bancker was president of the Franklin Insurance Company and had a passion for collecting scientific instruments, in particular those relating to optics. As noted in an 1859 issue of *Cosmos*: 'The scientific passion that suddenly seizes wealthy American merchants presents a truly wonderful side ... [Bancker] has been taken by such a great love for optics that he would not forgive himself for letting a new device go by without immediately acquiring it.'

Bancker was, it seems, in the habit of scouring the pages of scientific journals of the day, and buying up anything that interested him. His imagination was piqued by Scott de Martinville's ingenious device – indeed he may possibly have first heard about it in *Cosmos* itself, which ran a piece describing the invention in its Christmas 1857 issue. He had to wait a couple of years, however, before finally acquiring the first commercially available version of the phonautograph. We know all this because, following Bancker's death, his collection was auctioned off and from the auction catalogue we learn that the phonautograph and 30 'acoustic drawings' – presumably the phonautograms – ended up at the then recently founded Stevens Institute of Technology in Philadelphia.

Early sound scholar Patrick Feaster discovered these references to Bancker's collection while digging around in court case transcripts at the Thomas Edison National Historical Park in 2014. And in a mind-blowing coda to the subject, he realised they represented not only what would have been the earliest working phonautograph on American soil, but they also contained references to it being tried and tested. This would have been at the Stevens Institute sometime in the 1860s. In other words, these would have been the earliest sounds ever recorded in America.

Scott de Martinville never managed to read the sound waves as he hoped. But 150 years later, they were resurrected and played back by a team of scientists and audio historians, including Patrick Feaster, who scanned and processed the traced waveforms, turning them into digital audio files. That all sounds very simple, like any of us could have done it with a scanner and a copy of GarageBand. But it wasn't.

For a start, there's the problem of quality. If we could get pure, clear playback simply by scanning a boar's bristle tracing on paper with an app, why on earth would we still make vinyl? The answer, of course, is because these weren't designed for playback, and we can't get pure, clear tones from boar's bristle on paper. Through the lens of the modern audio groove, these tracings are almost laughably poor, violating many of the fundamental requirements of signal recording. Indeed, at first the cutting-edge software was completely stumped, failing to make any sense of the smeary scratchings and causing the team to worry that phonautograms might remain mute.

One problem they had to overcome was time. The number of vibrations per second decides pitch, but without a reference point, lines are meaningless. However, back in 1859, Scott de Martinville teamed up with a maker of scientific instruments named Rudolph

Koenig. In a bid to make the invention more useful as a means of study, Koenig had added the tuning fork as a kind of time signature control signal. This would be struck when the machine started recording. Both the stylus and tuning fork would therefore leave parallel lines – the tuning fork vibrating at its constant speed alongside the quivering stylus. This meant that someone studying the document could literally count the undulations carved in soot by the tuning fork to determine the speed of the recording.

It was these – the most advanced surviving phonautograms with their helpful tuning fork time key – that could be reconstructed. Audio historian David Giovannoni travelled extensively, gathering high-resolution copies of the documents, while lab scientists Carl Haber and Earl Cornell converted the images using modified software originally developed at Lawrence Berkeley National Laboratory. Following a long, drawn-out process of signal processing and manual time correction, the team went public, and in March 2008, the recording of a woman singing 'Au clair de la lune' on 9 April 1860 was heard all over the world. It sparked hilarity during a memorable segment on BBC Radio 4's normally staid news programme *Today*. The usually steadfast newsreader Charlotte Green was left in fits of giggles after broadcasting the clip of strange, quavering singing, after which someone had whispered in her headphones that it sounded like a bee in a bottle.

In the aftermath of the initial story, Giovannoni and his team, known as the First Sounds collective, were continuing to experiment with a new, improved playback approach, which essentially stopped treating the phonautographic tracings like a phonographic groove. This approach graphically converted the trace into a signal of varying width that can be read as an optical film soundtrack. This still couldn't deal with the most seriously malformed tracings, but it could enable them to unlock sounds from many of the surviving phonautograms. Plus, it allowed them to create a new manually corrected 'Au clair de la lune', and it was during this part of the project that they realised they had misread the speed of the recording. The new, more accurate version was slower, lower, more mournful, and recognisably a man. And, in all probability, it was the voice of Édouard-Léon himself.

He was far from the only one to be experimenting with the so-called graphical method of studying sound. In London, British engineer William Henry Barlow wrote to the Royal Academy in the 1870s, attempting to generate interest in a new gadget he called the Logograph. This was similar to Scott de Martinville's phonautograph, in that it was designed to record and study speech. An illustration printed later in *Practical Applications of Electricity* shows a disembodied mouth speaking into a simple trumpet-shaped instrument with a drum-like membrane at the far end. This would trigger a spring-mounted marker, which would pass over a moving strip of paper – essentially employing the same technology as telegraphy.

An important difference between this and the phonautograph was that, while the latter recorded sound waves, Barlow's machine was designed to record breath impulses – the idea had come to him while watching a pipe smoker and observing the pockets of smoke as he spoke. In other words, the membrane was designed to eliminate and filter out sound waves so that the instrument could pick up changes in air pressure.

Barlow wanted the machine to become a means of study. He foresaw some practical use in the worlds of science and business, and he too hoped that with practice, humans could learn to read the wavy lines the speech left in the paper. He saw it as a labour-saving device that would render the stenographer redundant, which explains the subtitle given to the machine in *Popular Science Review*: 'Writing by the Voice'. Barlow submitted examples of his read-outs from the machine, alongside the words that had produced the shapes, such as 'what are the wild waves saying?', from Charles Dickens's *Dombey and Son*, and his pleasingly Victorian test phrase 'the vibratory action is now very decided'.

This may seem a bit of a blind alley, unrelated to the coming of the phonograph. After all, no sound waves were recorded and so, unlike the phonautograph, however you analysed the logograph's tracings, you would never be able to reconstruct Barlow's voice. But his invention clearly shared attributes with Edison's Eureka moment (its conception borrowed from the world of telegraphy; it traced its markings on strips of paper), showing more than one person was experimenting along similar lines, using similar techniques, at around the same time.

THREE

Oddballs

'This was a marvellous piece of mechanism, though for some unaccountable reason it did not prove a success.'

P. T. Barnum

Preacher John Wesley visited Lurgan in County Armagh, Ireland, in 1756. There he met an ingenious man named William Miller, who had apparently made a fully working mechanical figure of an old man that stood in a case with a curtain drawn. Whenever the clock struck, the figure drew back the curtain, 'turned his head as if looking round the company and then said, with a clear, loud articulate voice "past one, two, three" and so on'.

That is more or less all we know about the possibility of an early speaking machine, or automaton, in eighteenth-century Ireland. But if true, it joins a parallel group of oddballs who were coming at the problem of sound from the total opposite direction to those working with phonographs, and attempting to synthesise the human voice by mechanical means.

This may seem unrelated at first glance but, trust me, it's worth it. For one, it's weird, and for another, it helps to understand the cultural impact of the first phonographs, the bizarre ways they were marketed, and to explore why they were originally dubbed 'talking machines'. The phonograph's early biographers all tended to include these speaking automatons in their chronological narratives because, to them at least, these inventions were close cousins. But if you're sick of all this back story, feel free to jump ahead to the next chapter, as that has a good bit about telegraphs.

Hungarian author and inventor Wolfgang von Kempelen is best remembered for The Turk – a supposed intelligent automaton consisting of a life-sized human head, with black beard and grey eyes, and seated behind a cabinet on top of which a chessboard was placed. The machine appeared to be able to beat very capable players, but in fact inside the large wooden cabinet was a hidden human chess player, who powered The Turk by use of an inverted chessboard and a complicated series of levers.

The elaborate subterfuge of this mechanical illusion would not be revealed for many years, after which time Kempelen and The Turk's subsequent owners had already toured the world several times over.

From the late 1760s Kempelen was working on another mechanical contrivance, in the form of a manually operated speech synthesizer.

Inspired by the vox humana, a reed stop common in pipe organs and given the name because of its resemblance to the human voice, Kempelen spent two decades working on his new machine, the most advanced being a monotonal model of the human vocal tract, operated by bellows, reeds, a rubber mouth and a string-operated wooden tongue. He recorded his work in minute detail in *The Mechanism of Human Speech, with a Description of a Speaking Machine*, which in turn inspired several other inventors to have a go at making their own. Meanwhile, a Swiss mathematician named Leonhard Euler had sent a letter to the Russian Academy of Sciences in St Petersburg. He challenged the scientific community to explain how it could be that such different sounds could be made by the flow of air through the vocal folds and tract, and speculated about building an instrument that could produce similar sounds or even understandable words. Soon after, the Academy announced a new prize to investigate this subject, which would eventually be won in 1780 by Christian Gottlieb Kratzenstein's 'vowel organ' – another vox humana-inspired contraption through which each pipe emulated a single vowel sound.

Magic and illusion were a staple entertainment for the Victorian concertgoer. Town and city music hall boards creaked with variety bill spectacles and standalone shows, while venues such as the Egyptian Hall in Piccadilly, London, became the home of residencies for Victorian magicians like John Nevil

Maskelyne and George Alfred Cooke. Mechanical marvels, magic lantern spectaculars and thrilling illusions, such as Pepper's Ghost,[1] trickled down from making newsworthy debuts to become, in simplified form, fairground and sideshow staples.

Out of this phantasmagoria came the Euphonia, or the Wonderful Talking Machine. This was the work of a strange and solitary man named Joseph Faber, who was born around 1800 and was schooled in Vienna, Austria. While recovering from a serious illness in his twenties, he got hold of a copy of Kempeler's *On the Mechanism of Human Speech* and caught the bug that would dominate the remainder of his life.

Faber's first working machine made its public debut in Vienna in 1840. Miffed by a perceived lack of interest, Faber destroyed his machine and moved to America. Once there, he began work on another, and a few years later there were reports of the new and improved version being unveiled in New York City. Again, however, the invention failed to make the impact he hoped for, and so he destroyed that one too.

Next, Faber unveiled a new incarnation at the Musical Fund Hall in Philadelphia in December 1845. By now, after two decades of trial and error, he had made

---

[1] A stage illusion named after its inventor, John Henry Pepper, which became a sensation in the 1860s. The technique employed a glass plate, set diagonally in relation to the audience. This could reflect a part of the stage not visible to the audience, where an actor could stand in spooky get-up. As the intensity of the lighting on the actor was increased, the reflection in the glass would become more visible, creating the effect of a ghostly apparition appearing in the middle of the stage, seemingly out of nowhere.

progress. His machine worked. During demonstrations, the bespectacled Faber would stand behind the device, operating 16 levers or keys, which between them projected 16 elementary sounds, powered by bellows. Perhaps unwisely, Faber had also given his machine a face; the levers controlled the movements of tongue, lips, jaw, and vocal cords, the latter shaped by the use of an all-important 17th key.

One fan of the spectacle was physicist Joseph Henry, the inventor of electromechanical relay, who was so delighted by the strange creation that he imagined a future where mass-marketed, robot-head speech synthesizers would be rigged up to telegraphs and made to deliver spoken messages received down the line. He envisaged the possibility of having a sermon delivered simultaneously to multiple church congregations by a series of wired-up, rubber-lipped speech-bots. Yet more success came to Faber the following year when the American showman Phineas Taylor Barnum got wind of the invention, renamed it the Euphonia, and took the machine and its inventor to London, where it became one of the staple attractions at the Egyptian Hall. Legend has it that the Duke of Wellington visited several times, convinced that the voice came from the exhibitor, until he had a go himself.

Sadly, it still did not set the world alight. Poor Faber, it seemed, was making one of humankind's first forays into the Uncanny Valley. While his talking machine was, by all accounts, the most life-like recreation of the human voice to date, it was also slow, mournful and creepy as all hell. Using the foot-operated bellows, Faber could by now make his breathy Eeyore sing a sepulchral rendition of 'God Save the Queen', but as often as it impressed some passers-by, it would provoke laughter and ridicule in others. In his autobiography, Barnum claimed the Duke of Wellington had tried it out and, after some instruction from Faber himself, managed to get the thing working, admitting it to be an 'extraordinary production of mechanical genius'. And yet, for some unaccountable reason, it did not prove successful. London theatre manager John Hollingshead described Faber as a sad-faced man, dressed in respectable but worn clothes, operating the keyboards slowly and deliberately. He suspected that the German inventor slept in the same room as his creation, and seemed to sense doom surrounding the professor and his 'child of infinite labour and unmeasurable sorrow'. Faber eventually disappeared from London, taking his marvel to the provinces, where it was even less appreciated. 'One day, in a dull matter-of-fact town, he destroyed himself and his figure.'

Despite this sad end, the Euphonia and its predecessors left their mark on the phonograph. Indeed, one devotee was a Scottish professor of speech named Alexander Melville Bell, whose young sons decided to tackle the problem themselves. And as one of these young sons would go on to invent the telephone and then help improve the ailing phonograph, it is perhaps worth drawing your attention to his first foray into sound reproduction.

The Bell brothers, Alexander and Melville, tried their hand at building a talking machine modelled after human vocal organs when they were aged 15 and 17 years old, and living at 13 South Charlotte Street,

Edinburgh. They divided up the work between them – Melville undertook the lungs and throat, Alexander the tongue and mouth. Melville knocked up an artificial larynx of tin and rubber, while Alexander made a palate, teeth, throat and nasal cavities from gutta-percha, a kind of natural thermoplastic in wide use at the time. He constructed a soft palate from a bag of rubber stuffed with cotton batting, and a wooden tongue with six or seven movable sections, also covered with layers of cotton batting. Sheets of soft rubber formed cheeks, and rubber bags hinged together to form lips.

In the initial grip of enthusiasm, they had planned to install bellow lungs and connect up all the movable parts to a kind of control keyboard. But by the end, they simply attached the tin larynx and blew through the rubber hole.

'Helped no doubt by our imagination we thought the apparatus produced a very good "Ah",' wrote Alexander. 'While my brother blew continuously through the larynx, I opened and closed the artificial lips with my hand. To our great delight the machine gave utterance, unmistakably, to the words: "Mamma! Mamma! Mamma!"'

They took their hellish contraption to the common stairwell in the building and were delighted when their imitation of an infant in distress actually provoked an overheard neighbour to ask whatever could be the matter with 'that baby'.

Talking of babies ...

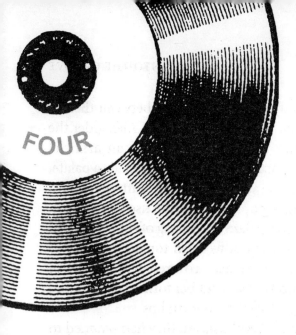

Telegraphs and Telephones

> '*I have invented a great many machines, but this [pats phonograph] is my baby, and I expect it to grow up and be a big fellow and support me in my old age.*'
>
> Thomas Edison

Unplayable paper records, creepy speaking machines, telegraphs and telephones together formed the crucible in which the phonograph was forged. Automatic telegraphs were recording messages via perforated strips of paper, while telephones showed us that sound waves striking a diaphragm at one end of a wire could then be reconstructed by a second diaphragm at the other end of a wire. Of the two, the telephone might seem the most obvious parent, but telegraphs had a stage-setting role to play and, if nothing else, formed Thomas Edison's lifeblood.

The first experimental telegraph was constructed in France, using 26 wires to pass messages between two rooms. Then, an English inventor built the first recognisable modern telegraph system, which sent signals along subterranean and overhead wires, each end connected to revolving dials marked with letters. In the 1830s, Samuel Morse and his assistant Alfred Vail developed an instrument that could record the messages it received using an electromagnetically controlled stylus that would emboss dots and dashes on moving paper tape. The first telegram in the United States was sent by Morse on 11 January 1838, across 2 miles of wire at Speedwell Ironworks near Morristown, New Jersey.

Soon, innovation was everywhere. Electrical telegraphs became the standard method for sending messages over long distances, and freshly laid submarine cables gradually connected countries and then continents. Railway companies tested systems to help resolve the whole 'how do we stop the trains hitting each other' question, while telegraph networks fought to up speeds and reduce costs. Telegraphs required skilled technicians on hand to read or write messages. Some of this skill could be removed by the likes of Charles Wheatstone's ABC system, which arranged letters of the alphabet around a clock face, so a signal would trigger a needle to indicate a letter. In Edinburgh, Scotland, a chemical telegraph marked out messages in readable dots and dashes. Forty miles away, in Glasgow, a teleprinter used compressed air to punch holes into tape. At the same time, telegraph networks in the United States were increasingly manned by sound operators trained to understand Morse aurally, thus doing away with the need for register and tape, and massively increasing the speed with which a message could be relayed.

With all this telegraph action in the air, suddenly we see how the telephone wasn't so far away. Certainly, the idea for the telephone was simple enough on paper: take a diaphragm, an electrified magnet and a spiral wire. As the diaphragm vibrates, it alters the current. These alterations of current travel to the other end of the wire, where they cause a receiving diaphragm to reproduce the sound.

Italian inventor Innocenzo Manzetti, another strange automaton obsessive who made a head-turning robot-like flute player, mooted the idea of a speaking telegraph in 1844. Then in 1861, Germany's Philipp Reis demonstrated an early proto-telephone, through which he succeeded in transmitting sound by electricity. In November 1865, *Le Petit Journal* described a demonstration of Manzetti's own telephone-like apparatus, the writer imagining a future where 'two merchants will be able to discuss their business instantly from London to Calcutta', remarking that the imperfect machine could already handle music and that, while the speech was recognisable as speech, only the more sonorous words had any hope of being recognised. Meanwhile, one Antonio Meucci set up a voice-communication link in his Staten Island home, connecting a second-floor bedroom (where his wife was ill) to his laboratory. In 1871 he filed a patent caveat – a kind of official 'I'm going to make this' notice – announcing his proposal for a Sound Telegraph, but he never managed to take it further. By the mid-1870s, Alexander Bell was well advanced with his new bi-directional gallows telephone, which transmitted electrical impulses between transmitter and receiver with identical membranes. It was able to transmit garbled sounds but was not yet good enough to transmit clear speech. Around the same time, the American engineer Elisha Gray was experimenting with liquid transmitters.

People still disagree about who was first to the punch out of these two. Bell was working in 292 Essex Street in Boston, Elisha Gray was a professor at Oberlin College. Gray applied for a caveat of the telephone on the same day Bell applied for his patent of the telephone, 14 February 1876. While their paths wouldn't have literally crossed on the day – lawyers did all the filing on their behalf – it's generally accepted that Bell's lawyer got to the patent office first. The unarguable fact is that the US Patent Office awarded Bell the first patent for the telephone, Patent Number 174,465, rather than honour Gray's caveat. And most importantly for our story, Bell had successfully proven that sound waves striking a diaphragm could be reconstituted by a second diaphragm on the receiving end.

Into this hotchpotch steps Edison, a man you can put anywhere on a scale with 'world's greatest inventor' at one end and 'skilled capitalist and self-promoter' at the other. Some like to downplay Edison's genius, stressing his indisputable track record of running with other people's ideas. But in the simplest terms Edison was a workaholic, motivated by money, yes, but only in the sense of the freedom it granted him to continue working. Even as a wealthy man, he preferred the camaraderie of the workshop and laboratory to the ballroom or tea party. And this clarity of purpose, this single-minded drive towards utility and profitability, is what set him apart from many a tinkering dreamer.

Edison grew up in Port Huron, Michigan. He was the seventh child of a schoolteacher called Nancy and a rebellious Dutch-Canadian called Samuel, who at various times had tried his hand at tailoring, roofing and barkeeping but later, when moderately famous simply for being Edison's father, would gleefully describe himself as 'master of smoking, drinking and gambling'. Nancy, on the other hand, was rather more on point, and Edison's self-belief, he said, was built on the foundations of confidence laid by his mother. More than one of Edison's siblings died in childhood, and a bout of scarlet fever meant that from age 12 he himself had severe hearing problems. However, he was by now an obsessive reader and experimenter who always wanted to know how things worked, and later he wrote that he felt his partial deafness helped him – allowing him to filter out distractions, to focus fully on his work.

Nancy[2] taught Thomas writing and arithmetic, as his chaotic brushes with formal education would amount to just a few months in total. It's easy to imbue a photograph of an individual with the characteristics you believe that person to have, but a commonly reproduced photo of the teenage Edison certainly shows a boy comfortable in his own skin, full of a chin-out self-confidence.

A teenage Thomas became a newsboy on the Grand Trunk Railway, riding the train between Port Huron

---

[2] Nancy died in 1871, when Edison was 24. Three weeks later, Samuel started a relationship with his 16-year-old housekeeper, Mary Sharlow, with whom he had three daughters. Samuel died in 1896 at the age of 92.

and Detroit. He expanded his business beyond selling the daily papers to selling candy and produce, even employing other boys on other trains, then, and this is the amazing bit, buying a printing press and producing his own modest newspaper – the *Grand Trunk Herald* – from the train's baggage hold, which he sold along with the other papers.

He invested his earnings in equipment and chemicals for conducting experiments. As the story goes, he wasn't above conducting these experiments in the baggage car, and when one of these caused an explosion, an angry conductor threw his printing press and chemicals off the train. Another turning point came in late 1862, when Edison was waiting at Mount Clemens, Michigan, and as a train approached he noticed a toddler playing on the tracks. He sprang forward and rescued the child, who turned out to be the son of the stationmaster, James Mackenzie. In gratitude, the Mackenzies fed him for several weeks, while James senior gave him a crash course in telegraphy, knowledge Edison soon put to work.

Edison became part of a nomadic community of skilled telegraph operators who would take short-term positions in offices across the country. However, unlike his peers, he wasn't satisfied to 'maintain, repair and adjust'; he wanted to make telegraphs work better. He was also a more gifted receiver (transcribing incoming signals) than sender (tapping out messages on a telegraph key), and began working for Western Union in Cincinnati, Ohio, spending long nights transcribing news transmissions. Working the night-owl shift left him with plenty of time to read and experiment during the day. He began sketching out ideas for new relays and repeaters that could increase the strength of incoming signals and transmit signals over greater distances and, most importantly of all, new multiple telegraph circuits that could send more than one message over the same wire.

Towards the tail-end of the 1860s, while working in Boston, Edison attracted investment for a new double transmitter, filed his first patent – for an electro-graphic vote recorder, and then filed his first truly successful patent – a simplified printing telegraph that could record stock prices. The original 'stock ticker' was invented by Edward Calahan in 1868. Edison improved the design with a mechanism that enabled all tickers on a line to synchronise, so that they simultaneously printed the same information. The transmission of real-time numbers from exchange floors to brokers and investors across the country completely altered the landscape of New York finance in the late nineteenth century.

By 1870 Edison had quit his job to become a full-time inventor, opening his first commercial workshop in Newark, New Jersey. Newark was a thriving manufacturing hub, and Edison surrounded himself with skilled engineers and workmen, many of them recent migrants from Europe. He made a string of innovations in telegraphy, culminating in the quadruplex telegraph circuit, also finding time to forge two important partnerships: the first to his wife, Mary Stilwell (in a notebook from February 1872, he wrote: 'Mrs Mary Edison My Wife Dearly Beloved Cannot invent worth a Damn!'); the second to Charles Batchelor, a Manchester-born draftsman and machinist who first crossed the Atlantic to New Jersey to install textile equipment and would become one of Edison's closest colleagues and confidants.

The phonograph was still seven years away when Edison signed another star player in John Kruesi. Kruesi was born in Heiden, Switzerland, grew up in an orphanage, served an apprenticeship with a locksmith, became a machinist in Zurich before stints in Paris, London, Belgium and the Netherlands, and finally arrived in the United States in 1870. He was working for the Singer Sewing Machine Company in New Jersey when he heard about an exciting new inventor in town, and soon Edison employed him in his workshop. Charles Batchelor described Kruesi as 'the most indefatigable worker in the crowd', able to problem solve and make things quickly.

In 1875, Edison took another step in our direction, towards the moment in history the phonograph was invented, by signing an agreement to work on acoustic telegraphy. This was funded by Western Union, who wanted something to compete with the newfangled acoustic system just invented by Elisha Gray – something that could transmit multiple messages by sending acoustic signals of different tones. This deal not only took Edison into the field of acoustics but also brought in enough money to relocate him to the birthplace of the phonograph – Menlo Park, New Jersey.

So here we are, in the company of a self-taught, ceaseless experimenter, happiest in autonomy and now with enough cash to enjoy it, always determined to see how things could be made better, surrounded by a crack team of engineers and mechanists, in a purpose-built innovation engine, and with his oil-stained fingers in two related pies – telegraphy and telephony.

Yet he's about to be pipped to the post by a poet.

FIVE

The Barb of a Feather

'This surface is a part of a disc to which is given a double movement of rotation and rectilinear progression.'

Charles Cros

Charles Cros was a romantic dreamer, a bronzed bohemian, a multi-talented, straggly-haired absinthe drinker, at the tail-end of a decade-long affair with a doyenne of Paris's literary and artistic scene. He was witty and funny, he spent time educating and entertaining deaf and mute children, apparently delighting them with the amazing contortions of his face, penned nonsense poems such as 'Le Hareng Saur' ('The Smoked Herring') and popularised poetic monologues – what could be seen as a precursor of stand-up. When he died at the age of 45 in 1888, the *South Wales Echo* ran his obituary under the headline 'Rhymer, Chemist, Fiddler and Buffoon', remarking that he touched every subject from politics to potatoes, calculus to cauliflowers. He became a regular at Le Chat Noir, the famous cabaret in Montmartre whose clientele included Claude Debussy and August Strindberg, and he played a key role in Les Hydropathes, an important Parisian literary group that emerged at the end of the nineteenth century. You could draw a comparison with the racing drivers James Hunt and Niki Lauda, where Cros was Hunt (free-spirited, naturally talented, a heavy drinker) and Edison was Lauda (intense, serious-minded).

His long-time partner Nina de Callias was a composer and pianist herself. She ran a salon in Paris where writers and artists would gather. Around the time of our story, she was dressing herself in a black chiffon ruffled dress and leaning back on a white chaise so Édouard Manet could paint her as the *Dame aux éventails* (*The Lady with Fans*, c.1874). Cros, while not globally famous, was already a cultish figure, a published poet, a person whose energy, artistic zeal and scientific proposals had made the popular press on several occasions. Today he is celebrated as a nearly-man of Parisian bohemia because what happened to him happened to him twice.

Just around the time Nina de Callias was arranging herself on the white chaise, Cros was appearing in papers on both sides of the Atlantic, attempting to get traction for an idea he'd had ahead of the coming Transit of Venus. This rare celestial event was due to occur on 9 December 1874, when the planet would pass between Earth and the Sun, and so be visible from Earth as a small black dot moving across the face of the Sun. This would provide an opportunity for improved measurements and observations, and various plans were already afoot to study the phenomenon from vantage points around the globe.

As reported on the front page of the *Lake County Star*, Michigan, Cros argued that this transit was a perfect opportunity to make contact with Venus – and any potential intelligent life forms.

'It is possible that Venus is inhabited,' he said, 'that among its inhabitants are astronomers; that the latter judge the passage of their planet across the solar disc to be an object of curiosity; finally, it is possible that these savants will strive in some way to make signals to us at the precise moment when they might suppose that many telescopes will be levelled at their planet.'

He was convinced pinpoints of light that had been observed on Mars and Venus were large cities (the pinpoints of light may have been high clouds illuminated by the Sun). He'd previously published a 'Study of the Means of Communicating With Other Planets' and petitioned the French government to build a giant mirror that could be used to communicate with Martians and Venusians. He even sent copies to the Académie des Sciences, and wrote a short story called 'Un Drame Interastral', depicting two young lovers, on Earth and Venus, sending mirror images of flowers to each other.

Then he almost invented colour photography. His idea was that a single scene could be photographed through coloured glass filters and the resulting negatives combined into one. This was similar to a prototype being hatched by French pioneer of photography Louis Ducos du Hauron. And on the same day, 7 May 1869, both men, independently and unknown to each other, presented their methods to the Société française de photographie. Cros ended up conceding the invention to Hauron, despite having sent a paper on the subject to the Academy of Sciences two years earlier. And amid all this, Cros published his first book of poems, *Le Coffret de santal*, most noted for his paean to absinthe, 'Lendemain'.

Cros's second nearlyvention came in 1877. How long he had been pushing around his ideas is unknown, but his track record of study in photography and telegraphy means it could have had a long gestation. Perhaps singed by his experience with colour photography, this time he wrote it all down and sent it in a sealed letter to the Académie des Sciences.

He titled the letter 'Process of Recording and of Reproducing Audible Phenomena'. And the most celebrated passage, the one that might give you goosebumps, runs as follows: 'A light stylus is connected with the centre of a vibrating membrane; it terminates in a point – metallic wire, the barb of a feather, *etc.* – which bears upon a surface blackened by a flame.'

So far, so phonautogram. Then:

'This surface is a part of a disc to which is given a double movement of rotation and rectilinear progression.'

Ahhh. Isn't that lovely? That sounds like a record.

He further explained that the membrane would leave undulating tracings within a spiral groove. The next step was that by 'photographic process', you'd convert that undulatory spiral to an intaglio print, in a 'resisting material like tempered steel, for instance'. So in other words, you'd make a kind of acetate from the original lampblack recording, then set that disc spinning at the same speed as the original was recorded. Then you'd set a metallic point, held by a spring, into the groove, and connect the end with the centre of a membrane adapted for sound reproduction.

So you see? That's why, because of his being the two-time nearly-man and a romantic, and because Edison has quite enough credits to his name already, and because of the fact that this did indeed come first and that it actually sounds much like a modern disc-shaped record, whereas Edison's first record was a weird tubey-type thing, Cros's achievements are widely and justly celebrated.

That's not to imply that his method was better than Edison's. For a start, the Cros method required a chemical engraving process, while Edison's was purely mechanical. However, whether Cros impresses you or not, this is the earliest written account of a procedure to retrieve recorded sound that might actually have worked. In the aftermath of the public unveiling of Edison's breakthrough, Cros asked the Academy to open his sealed packet to prove the priority of his concept. However, he did not have the resources to build a prototype. When he filed his own patent in the following months, there weren't any drawings to show how it would work. And it wasn't until relatively late in 1877 that he finally gave his invention a name – the 'paléophone', from the Greek *palaios* (ancient) and *phōnē* (voice). The ancient voice.

Cros's patent was granted on 1 May 1878,[3] with a term of 15 years, but aside from one addition granted on 3 August 1878, nothing more was done with it. However, it did clearly specify methods of sound recording in which he anticipated Edison, and left the Cros patent open for use in France.

Cros's achievements live on. He is well remembered in France, not least thanks to the lobbying of his friends and contacts, who spent much of the late 1870s arguing that he should have some official recognition. The Académie Charles Cros, founded in the 1940s, today gives out its annual Grand Prix du Disque, which recognises outstanding achievements in recorded music and musical scholarship. Cros remained an unprosperous bohemian to the last. His muse, who had a history of alcoholism and depression, died at the age of 39. And four years later, Charles went the same way. His friend, writer Alphonse Allais, spoke at his funeral. 'He immediately appeared to me as I always knew him, a being miraculously gifted in all respects, a strangely personal and charming poet, a true scholar, a disconcerting fanciful, a most reliable and good friend.'

---

[3] An Englishman named F. B. Fenby obtained a patent on 13 January 1863 for an 'Electro Magnetic Phonograph'. This was the first time the word 'phonograph' was used to describe a recording device. His machine would record the playing of a keyboard instrument on paper tape. The tape could then be used to play back the keyboard, mirroring the sequence as played by the recording artist – a bit like a Player Piano or Pianola. There is no record of a model of this device ever being built.

SIX

The First Phonograph

'Made phonograph today.'

Charles Batchelor's work diary, 4 December 1877

The phonograph started a lot of arguments, many of which really get going in the story in a few years' time, in a flurry of patent disagreements that played out in court. Happily the transcripts of sworn testimonies give us eyewitness accounts from the summer of 1877 – albeit described a long time after the fact – and there's lots of documentary evidence, too. Edison and his colleagues were in the habit of keeping detailed laboratory diaries, scrapbooks and worksheets, and while the archives become much more complete *after* 1877, there's still quite a bit for us to work from.

Edison is said to have been a man who epitomised energy, utility and efficiency in everything he did. It was important for him to feed this persona, as it was attractive to investors. A widely circulated version of his courtship with his first wife, Mary Stilwell, had him spotting her at work, popping the question soon after, Mary accepting the following day, and a week later they were married. This was contradicted by an earlier published interview with Mary in which she said she had never worked for Edison, they had met during a visit to his works in Newark, and that he courted her for AGES. There are, of course, many reasons why she might have preferred to keep her possible employment under wraps but, whether she was employed by Edison or not, this story illustrates that there is more than one version of what happened at Menlo Park in 1877, and that it's important always to keep in mind that Edison had a reputation to uphold and patents to protect.

In 1876, Edison sold up in Newark, moving his family and business to Menlo Park, New Jersey. His new base was a white two-storey building, about 30m (100ft) long, set on the crest of a hill. The ground floor had much of the heavy machinery, workshops, shelves and cases with models of experiments and inventions, a drafting room and a scientific library. The upper floors had a long, wide laboratory, with workbenches and shelves lined with thousands of bottles of chemicals.

The attributes Edison valued most in colleagues were precision and endurance. While the workforce would expand rapidly, for those first couple of years at Menlo there were about a dozen permanent staff, including Charles Batchelor, John Kruesi, Swiss machinist Charles

Wurth and an Irishman named James Adams. They were isolated enough to be free of the distractions of city life but close enough to the railroad to give them relatively quick and easy access to any resources they needed. Contemporary testimony paints a picture of a place where each worker was given autonomy; everyone on the team knew what problem the laboratory was tackling at any particular time and individuals were trusted to work through things on their own. There was no strict factory-like time-keeping; however, their hours were long. Shifts frequently lasted into the small hours, when Edison would sometimes treat workers to midnight snacks and entertainment, often playing an organ installed at the end of the laboratory. He grew to prefer night hours, as they were less likely to be interrupted by visitors. He never grew discouraged or downhearted when something didn't work, and his overpowering boundless zeal to find a solution inspired a loyal, can-do culture.

At work Edison was talkative. Outside, he would converse with anyone interested in his inventions but definitely preferred the company of those who understood how they worked. One writer described him as a pleasant-looking man, 5ft 10in tall, dark hair, slightly silvered, with wonderfully piercing grey eyes. For formal events he could be immaculately turned out, clean-shaven, in smart, well-pressed suits, but at the lab he was more likely to be dressed in acid-stained clothing, with dusty brows, discoloured hands and dishevelled hair.

Edison and the team were working on acoustic telegraph systems, which used sound tones to transmit messages. Following public demonstrations of Bell's telephone at the Philadelphia Centennial Exhibition in 1876, rapid development of a Bell alternative for Western Union became another major project for Menlo Park. By the summer of 1877, they had spent several months on telephonic tweaks and variants and had been experimenting with repeating telegraph messages. Similar to the Edison's stock ticker, the latest repeating device cut dots and dashes into a wax-covered disc, which could then be run under another instrument, from which the message could be transcribed or sent on to another. So, as you can see, the entire Menlo machine focused on sound, messages, repeating and recording.

One day, Edison was holding a telephone diaphragm mounted in a rubber mouthpiece in his hand. He was making various noises into the mouthpiece – some accounts have him singing – and feeling the vibration of the centre of the diaphragm with the tip of his finger. Then he turned to Charles Batchelor and said: 'Batch, if we had a point on this, we could make a record on some material which we could afterwards pull under the point, and it would give us the speech back.'

We're now tantalisingly close to the first record player.

In a few minutes, Batch had indeed fixed a point to the centre of the diaphragm, which was mounted onto a grooved piece of wood that had been used for an old automatic telegraph. Then strips of waxed paper, used for making condensers, were prepared. Batchelor pulled a strip through the groove while Edison shouted 'Halloo, Halloo!' On pulling the paper through a second time, they heard Edison's words played back to them.

While they might not have been able to make out specific words, it was recognisably the same sounds that had just been emitted from Edison's own mouth – faint but true. Edison wrote at the bottom of a laboratory sheet: 'Just tried experiment with a diaphragm having an embossing point and held against paraffin paper … the vibrations are indented nicely and there's no doubt that I shall be able to store up and reproduce automatically at any future time the human voice perfectly.'

Talking about it later, Batch didn't recall a sense of immediate excitement. In fact, it was more of a shrugged 'of course'. Everyone involved could see that it would work, that it should work, and so they were relatively unfazed to see that it did work. It took some time for the magnitude of what had happened to sink in. Still, there was enough excitement in the room for the team to try several modifications that same night – different thicknesses of paper, crimped paper, paper placed edgewise, metallic foils.

For several weeks they continued experimenting, trying out waxed paper strips, spirally grooved discs, diaphragms of different shapes, sizes and substances, long spools of paper tape, different widths of groove and depths of wax. The aim at this time was a method for recording telephone messages, although a laboratory sheet from November records Edison deciding that the new tech could be used in speaking clocks and talking dolls.

The earliest mention of tin foil comes from another sheet, dated 17 August. There, among various thoughts and ideas on the current state of the project, including the use of sealing wax, cork diaphragms and paraffin paper, is scrawled 'another idea', to indent paper in spiral grooves and cover this in tin foil, which the point would then 'easily indent'.

The first outsider to be let in on the secret was Edward H. Johnson, a former telegrapher and business associate, who was out demonstrating Edison machines. He sent a letter on 4 August saying that in some 'speechifying' in Philadelphia he had mentioned what Edison 'proposed to do in the way of recording speech'. Apparently the crowd response was immediate and enormous. Johnson became something of a thorn in the team's side, essentially saying: 'People are really excited. Make this happen!'

By September, they had prepared a kind of press release – one that never left the laboratory. Under the headline 'apparatus for recording automatically the human voice and reproducing the same at any future period', it trumpeted the possibility of recording speech on paper in a 'new, unexplored field of acoustics'.

The first official announcement came via a letter written by Johnson, with the OK from Edison, which was published in the 17 November edition of *Scientific American*. This was coupled with an editorial headlined 'A Wonderful Invention – Speech Capable of Indefinite Repetition from Acoustic Records'. It begins by predicting that the new invention will create the 'profoundest of sensations' and liveliest of emotions, as it allows humans to once more hear the 'familiar voices of the dead'. Then the letter from Johnson describes the breakthrough in more detail. In fact, what it describes is essentially a strip phonograph – a design where a long

THE FIRST PHONOGRAPH

continuous strip of paper wrapped around a clockwork cylinder is drawn beneath an indenting point.

The word was out. Papers and publications began to run stories on what was coming. But while an outsider in the know would have been expecting some kind of strip phonograph, the second-floor lab team had now pivoted away, towards a foil, grooved cylinder and hand-cranked design. Cylinder and foil did away with any of the problems caused by feeding strips of paper into machinery or by speed variation set by a spiral disc design.

A sketch was handed to John Kruesi on 29 November, and the first prototype appeared six days later. This had two diaphragms mounted on either side of a finely grooved brass cylinder – an angled mouthpiece of brass and gutta-percha for recording, a second for listening, the metal stylus points mounted on fine watch springs. (The second mouthpiece was soon abandoned when they realised you could use one for both recording and listening.) Edison wrapped the brass cylinder in tin foil and began cranking the machine, while saying the children's nursery rhyme 'Mary Had a Little Lamb' into the recording diaphragm. He pulled back the recording point, reversed the crank back to the start, pressed the reproducing point to the tin-foil groove and turned the crank again.

It worked.

Now Menlo Park *was* buzzing. Edison's oldest daughter, Marion, loved the laboratory and marvelling at all the weird contraptions as she brought up her father's lunch. She recalled visiting the day after the phonograph had been made. The place was largely deserted as everyone was catching up on sleep after a long shift. There was a worker who showed her the device. When she heard her father's voice, she jumped up and down with excitement.

Finally they had a proper working prototype, and they wasted no time. As noted above, the rumours and reports of speech-recording technology doing the rounds were short on detail and still described a kind of strip-paper message taker. Then, in December, the team paid a visit to the offices of *Scientific American*, placed the new machine on the desk, cranked its handle, and watched as a pre-recorded message asked open-mouthed magazine staff how they were and what they thought of the phonograph.[4]

Aspects of this discovery story changed over the years. The shorter accounts gave the impression that the jump from diaphragm-on-finger musings to finished machine was virtually instantaneous. Some had Edison inspired not by a telephone diaphragm at all but by a kind of voice-operated toy. Edison himself described working on a telegraph repeater, which was indenting morse on paper, and when he ran the paper very rapidly through the machine, it gave off a humming sound that reminded him of human speech.

---

[4] The word phonograph had been used before. It had also been a used for a patented device before. But it hadn't been used for a recording device before. Or had it? Abbé Lenoir, a clergyman-cum-scientist-cum-science writer, who had been writing about cutting-edge acoustics and optics for years, described Cros's process in an article published in October 1877. And in that article he called Cros's paléophone a 'phonographe', describing how the device would give the user 'photographs of the voice'.

The 'prick on the finger' story, though, is neat. It can be summarised in just a few sentences and makes sense to the layman.

Another version of the story has an assistant betting Edison $2 that Kruesi's prototype would never work. In another it was a box of cigars, another a barrel of apples, another it was a bet between two assistants. The individuals involved also differed. One assistant in an earlier version was James Adams, but later he becomes John Kruesi. It could be that Adams made the bet, and to simplify the retelling he morphed into Kruesi. Some stories had Kruesi betting against himself, predicting that it wouldn't work, and in others he didn't even know what it was supposed to do. All accounts agree that a bet was won, possibly by Edison, and there was a feeling of uneasy jubilation at how well the prototype worked. Edison was always suspicious of things that worked first time.

Whether that first tin-foil test really was 'Mary Had a Little Lamb', or something else, has also changed. Certainly 'Mary Had a Little Lamb' was the standard verse that Menlo Parkers yelled into mouthpieces and diaphragms whenever they were conducting experiments in sound and telephony. I'd also like to add that anyone, making anything, tests as they go. Kruesi wouldn't hand his boss a machine unless he was pretty sure it worked, would he? If we believe that, it means that, while Edison's recording is remembered and accepted as the famous world-first recording and playback, it was, in all likelihood, preceded by some of Kruesi's own speeches in foil that have been lost to history. Maybe.

Another thing that has left historians scratching their heads is the timeline. Why was there such a long gap from that original discovery of the ability to record and reproduce sound (July) to the creation of a working prototype (December) and the filing of the first US patent (Christmas Eve 1877)? This has been further muddied over the years by the supposed original sketch handed to Kruesi, which had the handwritten instruction 'Kruesi – Make this' and was dated 12 August 1877. This date just didn't tally with various lab sheets and work diaries, and it was eventually discovered it had been added later, either on purpose to fabricate a timeline or in error as part of the Menlo archiving system.

The long gap, it seems to an idle outsider, could just be down to lots going on. There must have been some frustration that they hadn't got to the telephone first – they were so close, after all. And the fact that Western Union wanted them to come up with their own, non-patent-infringing telephone must have been very high priority.

Another interpretation, one that has been made before, is that Edison and his team really didn't know what they had. It was Johnson's enthusiasm, shared by the editor of *Scientific American*, that helped to galvanise the phonograph project. It's certainly interesting that right up until November, when that first letter appeared in *Scientific American*, they were still talking about a strip-paper design. Perhaps it was the enthusiastic reaction to the letter and editorial that pushed them to hurry up. Perhaps they were beset with fear, worried that it had been a mistake to let the secret out before they had a functioning prototype and a filed patent. The result of all this was a run of frenzied all-nighters, and a machine that was very different from what had been described in the previews.

The excitement afterwards is palpable in documentary evidence. In a letter to George Bliss, an associate pushing the Edison electric pen business, Batch wrote: 'You probably remember when you were down here about Edison's idea of recording the human voice and afterwards reproducing it. Well we have done it and have today shown it in New York to the *Scientific American* people who are now sketching the apparatus for a future issue.'

The wonder of the age had arrived.[5]

---

[5] The original phonograph went on tour. Then, in 1880, Edison gave it to a representative of the London patent office, and it remained in London for nearly half a century, finally returning home in 1928. Detailed blueprints were made, forming the basis for all sorts of displays and authentic replicas. The more gleefully geeky among you may like to notice, and to point out to as many interested parties as you can gather, that most of these replicas miss out the short-lived second mouthpiece.

SEVEN

Fade Away

> 'Friends at a distance will then send to each other phonograph letters, which will talk at any time in the friend's voice when put upon the instrument. How startling also it will be to reproduce and hear at pleasure the voice of the dead! All of these things are to be common, everyday experiences within a few years.'
>
> James Baird McClure, 1879

The Wonder of the age turned out to be really annoying. Edison was soon more profitably engaged with light bulbs, and the phonograph went into a decade-long torpor. That's the short version of what happened next. The long version goes on for ages, and includes some tedious stuff about patents. The perfect mid-length version goes like this ...

Edison and the team were suddenly busy even by their standards. Now that the phonograph had made its debut, they were knocking up technical drawings and descriptions, working on designs and prototypes that could help bind future developments to them, and, at the same time, marketing, selling rights and trying to make the darn thing more usable.

One of the first people to take the phonograph out on tour was Edward H. Johnson. Following the phonograph's debut at *Scientific American*, Johnson was loaned the original machine so he could begin to show off just what it could do. He soon penned a letter to Edison, urging the team to work on a more reliable exhibition phonograph. In particular he was keen they try to make a clockwork version, as this would do away with the difficulty of maintaining a steady speed.

Think about it: you had to not only maintain a steady speed while recording, but you also then had to try to match that same steady speed while playing back. Any errors, in either recording or reproduction, would result in wild fluctuations of pitch. And while this could and would be used to comic effect to delight and entertain, it could also derail the more serious demonstrations aimed at the business community.

With public interest in the new talking machine at something of a fever pitch, the Menlo Parkers produced a stop-gap phonograph – not much better but certainly a little better than the original. They ignored Johnson's calls for a clockwork motor, instead producing a larger machine, with a nice heavy flywheel and a larger recording cylinder, which together made relatively steady speed a bit easier. The addition of a reproducing horn also made the machine considerably louder and consequently much more suited to exhibitions and public performances.

In early 1878 they made several hundred of these improved machines and sent them out across the country and overseas. Within weeks stories and eyewitness accounts flooded the newsstands. The March edition of *Harper's Weekly* printed an illustrated report from a public demo in New York. The enthusiasm in the piece is palpable and, like many of the early accounts, it lends the machine character. 'This little machine …' will do anything asked of it, like a faithful confidant. It will 'talk, sing, whistle, cough, sneeze', along with any other 'acoustic feat'. There's a true sense of wonder at a machine that seems like magic, like witchcraft. There's absolute, unimpeded fascination at how it records everything, EVERYTHING so perfectly. It stresses this point several times: that every word is captured, exactly as it is said – an exact replica. 'In order to reproduce the words – that is, to make the machine talk – the cylinder is turned back …'

They're surprised at how simple it appears. Expecting complicated levers, pulleys and weights, such as might be seen in Faber's weird old speech synthesizer that was still on display at Barnum's, the phonograph seems modest, positively nondescript. They remark that it manages to successfully hear and talk without any of the fussy, over-engineered gubbins going on in our own vocal cords and eardrums.

This, then, is at the heart of why phonographs, for generations to come, would be known as Talking

Machines. This very literal approach to naming, echoed a few years hence by the coming of 'Movies', captures something of the simple amazement and wonder that was felt on encountering a machine that could replicate the human voice.

The report ends: 'There is no reason why we should not have all the great men of the age, as well as all the brilliant singers and actresses ... Let them sing or speak once in any place, their words and tones will be captured by the phonograph. The tin foil, whereon all they have said is duly recorded, will be electrotyped, and copies sold at so much a piece. We shall all waste a portion of our substance on this little instrument; and then we have only to turn a crank, or set a kind of clock-work in motion, in order at any time to hear the great ones of the earth discourse in our own parlours.'

For now, Edison split the commercial rights into three – the phonograph itself, dolls and music boxes, and timepieces. When the Edison Speaking Phonograph Company was established, he received $10,000 for the manufacturing and sales rights, as well as 20 per cent of profits. For his side, he had to agree to keep working on it. The rights to make dolls and music boxes ended up with an electrotyper who managed to miniaturise the phonograph in a series of prototypes, none of which ever went into production. Phonograph clocks and watches were acquired and worked on both at Edison HQ and in a base in Ansonia, Connecticut. The interesting geeky fact to note about these machines, which used cylinder records of sheet copper, is that these were the very first cylinders designed and developed purely to reproduce sounds, rather than to record and reproduce.

In June, Edison presented a list of the possible uses for the phonograph to the readers of the *North American Review*:

1. Letter writing and all kinds of dictation without the aid of a stenographer.
2. Phonographic books, which will speak to blind people without effort on their part.
3. The teaching of elocution.
4. Reproduction of music.
5. The 'Family Record' – a registry of sayings, reminiscences, *etc.*, by members of a family in their own voices, and of the last words of dying persons.
6. Music boxes and toys.
7. Clocks that should announce in articulate speech the time for going home, going to meals, *etc.*
8. The preservation of languages by exact reproduction of the manner of pronouncing.
9. Educational purposes, such as preserving the explanations made by a teacher, so that the pupil can refer to them at any moment, and spelling or other lessons placed upon the phonograph for convenience in committing to memory.
10. Connection with the telephone, so as to make that instrument an auxiliary in the transmission of permanent and invaluable records, instead of being the recipient of momentary and fleeting communication.

You'll notice what served as the initial spark for the phonograph – a means for recording telephone messages – is now tenth on his list, that the reproduction of music comes in at a modest fourth position, and his belief that it will be of use to the business community is top of the list. He also had high hopes that it would prove useful in medicine – James Baird McClure wrote how a leading medical journal had reported on the phonograph opening up vistas of medical possibilities too 'delightful to contemplate'.

> Who can fail to make the nice distinctions between every form of bronchial and pulmonary rale, percussion, succussion, and friction sounds, surgical crepitus, fatal and placental murmurs, and arterial and aneurysmal bruit, when each can be produced at will, amplified to any desired extent, in the study, the amphitheatre, the office, and the hospital? The lecturer of the future will teach more effectively with this instrument than by the mouth. The phonograph will record the frequency and characteristics of respiratory and muscular movements, decide as to the age and sex of the fetus in utero, and differentiate pneumonia from phthisis. It will reproduce the sob of hysteria, the sigh of melancholia ... the cry of the puerperal women in the different stages of labor. It will interpret for the speechless infant, the moans and cries of tubercular meningitis, ear-ache, and intestinal colic. It will furnish the ring of whooping-cough and the hack of the consumptive ...

With so much excitement over this new ability to record sound, it seemed that, for now, profit could be made from the tin-foil phonograph as a pure novelty item. While later court testimony claims Edison never considered it good enough for proper commercial development, he was for the moment contractually obliged to try.

Tin-foil records were, for all intents and purposes, single-use items. While you could play a recording more than once before it wore out, the act of removing the foil pretty much always destroyed that recording. Looking for a medium that did a better job of reproducing consonants, they tried out wax, but decided that when it came to reproducing it would only manage one playback. They worked on disc[6] machines, too, testing ways to increase the volume, and tried out various means of speed control, spending considerable time and money on clockwork motors, before eventually calling a halt to focus on other things.

The operating instructions supplied with those first exhibition phonographs were intimidatingly long, with frequent use of doom-laden phrases such as 'careful attention must be given to', 'particular care must be taken' and 'equal care must be observed'. One section admits: 'There is considerable knack in the effective use of the voice ... a good voice is sometimes rendered ineffective.' It starts by describing the application of foil, warning that improperly placed foil is liable to tear, and the embossing point to break. The 'Foil Wedge', too, has to be pressed firmly into the grooves, but with no part touching the embossing point, or the embossing point

---

[6] The really shorthand, over-simplified version of how records came about is that Edison invented the phonograph cylinder, then someone else came along and invented the disc. While this does tell a story, it's not strictly accurate. The initial invention was the *ability to record*. Edison's patents included both cylinder and disc-shaped records, and he worked on prototypes of both. He preferred the cylinder and so focused on that. The cylinder *caught on* first, but the idea of the cylinder and disc records appeared simultaneously.

will break. Towards the end there's a long list of the many destructible parts and what to do when they destruct. And, as if suddenly filled with overwhelming self-loathing, the instructions at this point throw up their hands and admit: 'No elaborateness of instruction, will be sufficient to enable an operator to become an expert immediately in handling the simplest mechanism.'

The fact that foil was a one-time deal was often played down in some of Johnson's advertising copy. Rebranded the 'Parlour Speaking Phonograph', it was now being dressed up as a novelty item for the drawing room. 'It talks. It whispers. It sings. It laughs. It cries. It coughs. It whistles. It records and reproduces at pleasure all musical sounds.' The print explains that, with work on adapting the machine for practical business use still ongoing, Edison and co. decided to sell the exhibition version for the public to enjoy. It then claims that the foil, which should hold '150 to 200 words', can be removed and replaced at any future time, thereby 'reproducing the same sounds that have been laid upon it'.

Only a handful of the tin-foil recording sheets are known to have survived in anything like playable condition. One made on a phonograph in St Louis, Missouri, on 22 June 1878 was digitally resurrected in 2012. It was found to contain a 23-second cornet solo, and a voice reciting 'Mary Had a Little Lamb' and 'Old Mother Hubbard'. The voice is thought to be that of Thomas Mason, a local newspaperman who had bought tin-foil phonograph serial No 8 for $95.50 in April 1878 and gave a public performance of his phonograph later that summer.

Earlier that year, in mid-January, the British public had heard about the new contrivance through an article in *The Times* and a Charles Batchelor letter printed in *English Mechanic*. *The Times* got wind of it from Henry Edmunds, a British engineer who had been on a study tour of the United States and had witnessed a very early demonstration. Engineer William Preece[7] immediately arranged for a new phonograph to be constructed, which was overseen by Edmunds.

This first phonograph on British soil appeared in public at the Royal Institution on 1 February, the grand finale to a lecture on telephony. By the time of the second appearance, at a Society of Telegraph Engineers meeting at the end of that month, Preece had three machines – a copy made by an amateur engineer, the improved Edison model with the bigger flywheel, and a third sporting a gravity-powered motor.

Tin foil made similar debuts across the globe. On a Thursday evening in August, members of the Royal Society of Victoria in Melbourne, Australia, were given demonstrations of several new gadgets. 'Among these the most interesting, perhaps,' read the *Argus* report the following day, 'was the trial made by Mr Sutherland with the phonograph, which was most amusing. Several trials were made, and were all more or less successful. "Rule Britannia" was distinctly repeated, but great laughter was caused by the repetition of the convivial song of "He's a Jolly Good Fellow", which sounded as if it was being sung by an old man of 80 with very cracked voice.'

---

[7] A major figure in telegraphy in Great Britain, who helped set up the Post Office's telegraphic system. He remained at the coal face of innovation for decades, helping to further both the fledgling telephone system and later wireless telegraphy, championing the work of Guglielmo Marconi.

FADE AWAY

The history of competing patents can be tedious. However, it is important to understand something that went against Edison at this point. With patents, you can add caveats. After you patent the initial device, you add to it, with innovations and variations on a theme. In so doing, you attempt to protect not only your initial invention but also the rights to its exploitation in a number of related forms. So in other words, you try to think up all the novel ways it could be used, and get there first.

That same year, final specifications for the British patent were filed[8] – including a disc phonograph. This wasn't just a case of saying 'you know what, we could make a disc-shaped one!' They had to actually design and build working prototypes and submit detailed technical drawings. For some reason, however, there was a foul-up between the filing of the British application and the American application. The result of this was that under US patent law, those extra devices and caveats that were now covered in Britain could not be protected in the United States.

With Edison's energy increasingly focused on electricity and the light bulb, virtually all experimental work on the phonograph ceased. The Edison Speaking Phonograph Company had to recoup investment, so expensive experimentation was discouraged and the team simply made more phonographs when orders came in.

---

[8] In Britain, the rights to exploit the Edison patent were held by the London Stereoscopic Company. A dealer catalogue from 1886 lists three models on sale: hand-cranked, no flywheel costing £5; a flywheel version at £10 10s; a third driven by a falling weight at £25.

A visit from a reporter from *The World* magazine, the results of which were printed in early September 1879, provides a stark contrast with the excitement of 18 months before. The reporter asks about the current state of the phonograph business.
We produce them to meet demand.
So how many is that?
About eight to fifteen a month.
Any improvements made to the phonograph recently?
Not specially.
Has Edison finished a disc phonograph capable of containing an entire sensational novel?
No. He's abandoned that idea.
Are they making clockwork phonographs?
No.
There's a short summary of the improvements that had been made since its debut: the grooves are finer, the diaphragm is mica, the needle is chiselled steel. But the piece goes on to explain that Johnson, who was intending to carry out further improvements himself, is currently in England introducing Edison's telephone. It ends with this depressing but thankfully inaccurate quote: 'I don't believe the phonograph will ever be much further improved.'

The following year, *The Philadelphia Record* paid a visit to the Edison laboratory to report on a workplace where everyone and everything was focused on the electric light. The phonograph is seen abandoned on a table in the laboratory, surrounded by other contraptions, covered with dust, looking dilapidated and neglected, the needle broken. Edison points at it and says 'that's the phonograph', and moves on without further comment.

EIGHT

The Volta Lab

'I am a Graphophone and my mother was a Phonograph.'

Alexander Melville Bell

The Second major step in recorded sound was the graphophone. And the slightly odd thing about the graphophone was that it looked and worked roughly like the original phonograph and was all but designed as early as 1881, but it didn't appear for ages and, when it finally did, it was sold as a dictating machine.

Public interest in the phonograph was dwindling and business suffering. Looking back on his invention, Edison told readers of *Electrical World*: 'It weighed about 100 pounds; it cost a mint of money to make; no one but an expert could get anything intelligible back from it ...' In some ways, it's akin to what happened with mobile telephones – the price tag, over-the-shoulder battery packs and lack of infrastructure meant they remained a fringe item for a generation. The tinfoil phonograph, briefly an awe-inspiring wonder, was now an unreliable novelty item. One writer quipped that people spent a lot more time talking about talking machines than talking into them. So Edison dropped the phonograph, leaving a void, into which stepped Alexander Graham Bell.

Bell was drawn to the phonograph, as it was very much in his field of expertise. In hindsight, he expressed surprise that he hadn't come up with the phonograph before Edison. It seemed such an obvious, natural step, and he was frustrated with himself for having not taken it. That might sound like someone barking 'I knew that one' at a gameshow host who's just revealed the answer to a question, but it's obviously true that the two inventors were exploring similar avenues, with similar technology at their disposal. While Bell certainly had respect for the man who got there first, he also felt it was an opportunity missed. This feeling was shared by Edison, who might easily have got to the telephone before Bell, had things played out differently. In the aftermath, of course, Western Union paid Edison to come up with his own, non-infringing telephone, and now it was Bell's turn to see if he could improve upon Edison's phonograph. Another driver might have been Bell's father-in-law. Gardiner Greene Hubbard was a stockholder in the Edison Speaking Phonograph Company, and was growing ever more frustrated that the returns on his investment were so disappointing and that Edison had stopped trying to make the phonograph any better.

THE VOLTA LAB

Despite being first to the telephone, Bell must have been aware of Edison's shadow. Light bulb aside, Edison's patent on behalf of Western Electric, the carbon button transmitter, was already being put to use in newly set up telephone exchanges. In response, Bell came up with a photophone transmitter. This worked using a light-sensitive selenium cell. The voice of the speaker would vibrate a diaphragm, which had a tiny mirror attached, which would then send a variable beam of light at the cell. Compare that last sentence with the words 'carbon button'; this gives you a clue as to why the photophone, though undoubtedly ingenious and ahead of its time, was not an immediate success. It was just too costly and complicated. And this pattern would be repeated. Bell's team tested out all sorts of ingenious ideas that often proved impractical for commercial development. They secured a handful of telephonic patents at the start of the 1800s, but none was successful. Edison, on the other hand, kept churning out telephonic patents that were.

In 1879, Bell set up a laboratory in Washington, DC, with Charles Sumner Tainter. Tainter, like Edison, was a mostly self-educated, hands-on character, inspired by years of devouring copies of *Scientific American*. Unlike Edison, he was an abstemious tea-drinker, with a great fear of public speaking. He had worked for several electrical companies in Boston, then joined a firm making telescopes and optics, before going it alone as a maker of scientific instruments.[9]

---

[9] Tainter also has an interesting connection with Charles Cros (who was so convinced humans should watch for signals from any Venusian scientists) in that he had worked on a US expedition to observe the Transit of Venus.

One of his clients was Bell. Tainter found Bell to have a 'strong magnetic temperament', and their association was pleasant and harmonious from the start. Bell invited him to join a new venture to experiment with the transmission of sound, and Tainter signed up, agreeing a $15-a-week salary plus a 10 per cent interest in any inventions that came from the partnership.

The lab opened on L Street, between 13th and 14th Streets, soon moving to 1221 Connecticut Avenue. There was not much capital to begin with, then on a trip to Europe, Bell was awarded a 50,000-franc Volta Prize for the invention of the telephone. This windfall funded much of the duo's work and, having been joined by Bell's cousin, a chemist named Chichester Bell, they called themselves Volta Laboratory Associates.

Right from the start, while also perfecting that photophone thingy, they began experimenting with the phonograph. Indeed, old man Hubbard lent them his own phonograph, and from trying it out, they quickly decided tin foil was simply not up to the job. One of the first tests they conducted involved covering the Edison machine with beeswax and seeing how well that worked. They soon decided that paper, in some form, would function better than foil. But even though they hit upon the wax solution fairly early on, they continued working, partly because they had to tread carefully through the forest of patents and partly to see if they could find something better. One experimental prototype reproduced sound from grooves and undulations by a jet of compressed air, another used liquid. 'One advantage of the arrangement,' Tainter wrote, 'is that wear on the record is or may be thus reduced, if not practically avoided. It may also be observed that this method ... enables sounds of considerable loudness to be obtained.'

When trying out disc designs, they attempted to head off drawbacks foreseen by both Cros and Edison, namely that when a record rotates, as the stylus gravitates towards the centre, the amount of groove being covered during a single revolution changes. Sound waves of a certain pitch leave traces that are more bunched and become more difficult to reproduce. So they worked on designs to overcome this problem, a machine that helped the stylus move along the groove at a constant velocity.

One famous first came early in 1881. Until then, all phonographs used 'hill-and-dale' recording, in which the stylus cuts up and down, perpendicular to the recording surface. With 'lateral cut', as the name implies, the stylus imprints sound from side to side, in a groove that remains a uniform depth. From Tainter's notes, we know that even before Chichester joined the fray, they had already designed and built a lateral-cut disc record, recognising the advantages that both sides of the disc could be used, and that a resulting disc could be duplicated via electroplating or a plaster of Paris casting.

In 1881, their work was briefly interrupted by an attempt to locate a bullet in President Andrew Garfield, who had just been shot twice by Charles J. Guiteau.[10]

---

[10] The attempted assassination of President Garfield temporarily turned the team's attention to the development of an induction balance to locate the bullet in his body – a kind of metal detector. However, the combination of bedsprings and an overbearing

Towards the end of that summer, they were making good progress, and began to worry about Edison, or indeed anyone else, getting wind of their researches before they had filed any patents. So, just as Cros had sent that sealed letter in Paris, they prepared a summary of their work to date, popped it all in a tin box, sealed the box, and placed it in a vault at the Smithsonian.

Fifty years later the box was opened. It was found to contain an Edison tin-foil phonograph with a cylinder coated in wax, on which was recorded Bell's father, Professor Alexander Melville Bell, saying: 'I am a Graphophone and my mother was a Phonograph.' The name graphophone, Bell argued, was more than a simple mischievous reversal of phonograph. He argued that any machine while recording was a phonograph, and when playing back it became a graphophone – a 'writing-speaker'. And as they considered their machine superior at playing back, graphophone was the correct name. Alongside the machine were current newspapers to prove the date, as well as testimonials, detailed descriptions of their work, including the origin of the word 'graphophone', and an electrotype of a zigzag, lateral-cut phonogram.

To modern sound historians, this strange, at the time unplayable electrotype disc is perhaps the most intriguing object within the Smithsonian box. They had succeeded in creating it just a few days before. Tainter wrote that, while the wax cylinder placed in the

---

presiding doctor, who thought he knew where the bullet was, foiled their device. The president died from infections to his wounds on 19 September, ending these experiments.

Smithsonian package was 'hill and dale' (or, as they called it, 'Edisonian'), this electrotype's vibrations were parallel to the surface: 'Thus forming a groove of uniform depth, but of wavy character.' Tainter knew this design had all sorts of theoretical advantages, but he had struggled to create any kind of player that could play back work without tearing through the groove walls. So, while they had been successful in creating this disc, they hadn't managed to play it back. And in terms of their long-term goals, this is why they kept returning to the hill-and-dale method.

One parallel I'd quickly like to draw at this point is with the Voyager Golden Records, launched into space in 1977. One fun fact is that NASA sent the records off to deep space without any means of playing them. They figured, logically enough, that any aliens clever enough to find them would have to be space-faring creatures and therefore have intelligence enough to build their own record player. Similarly, the sealed Smithsonian box included this alien electrotype, but was without the means of playing it.

This is echoed by lots of the surviving Volta material, most of which is now preserved at the Library of Congress in the US capital. This huge collection contains hundreds of offcuts, discs, cylinders and foils made between 1881 and 1885, some of which were recorded and never played back. Many of Tainter's experiments would have gone that way; an experiment showing that some kind of impression was made in a recording surface, and that sound waves had been preserved and could be seen, was a successful experiment even if they had no means to hear how it sounded.

All that changed in the mid-2000s with IRENE, a project to unlock historical grooved audio without touching the surface. IRENE, a partnership between the Smithsonian Institution, the Lawrence Berkeley National Laboratory and the Library of Congress, works by taking a kind of 3D topographic photograph of the surface of a record, then digitally reconstructing it. The title IRENE is a reference to the first piece of test audio that was produced using this imaging process – 'Goodnight, Irene' performed by The Weavers on a 78rpm shellac disc – and stands for Image Reconstruct Erase Noise Etc.

The project allowed scientists to not only preserve discs and cylinders without playing them but also piece back together broken or fragmentary records. Most importantly for us, however, the machine can also tackle all but unplayable formats that were produced at the hands of Bell and Tainter, allowing us to eavesdrop on the Volta Lab in Washington, DC, working at the start of the 1880s. They're freely available now, and I urge you to take a pause from reading this excellent little book to go and listen.[11] They are utterly strange and completely spellbinding.

You can hear Bell senior saying, 'I am a Graphophone and my mother was a Phonograph.' You can hear them reading out poetry and repeating names and addresses. They're not above trilling and whistling and singing. They were also in the habit of transcribing exactly what was said, or rather testing to an agreed script, so that they could study word-for-word impressions in the groove. There's one long recording where they read out

---

[11] See firstsounds.org/sounds/volta/.

THE VOLTA LAB

lots of numbers – perhaps to see how well their machine would work for business figures or the stock market. They frequently say words like 'barometer' and 'potato', or rather 'Poh - Tay - Toe!', as the strong Ps and Ts must have made good, visible impressions on the record. Test phrases include the old favourite 'Mary Had a Little Lamb' and their own Volta special: 'How is this for high!'

One of the first recordings unlocked by IRENE was an experimental glass disc dated 17 November 1884, on which the Volta team had managed to successfully record themselves repeating the word 'barometer'. With this disc, the diaphragm vibrated a wide stylus, which in turn allowed light to pass through a variable narrow slit. The light then left its trace on a rotating glass photographic plate. That sounds complicated but in fact was simpler than an earlier glass experiment that they described in the lab notes as resulting in 'a remarkably beautiful photograph … of a beam of light varied by transmission … on a glass plate in front of a slot, by a bichromate of potash jet, to which I recited Moore's "Believe me, of all those endearing young charms".'

Another glass disc contains what is possibly the first recorded swear word but is certainly the first recorded sound of human disappointment. This disc contains recorded speech, a gap, then a second portion of speech. The problem occurred at some point during the first section of the recording. It's not known what went wrong – it may have been a technical glitch or perhaps the speaker messed up an agreed part of their test script – but something certainly did, leaving behind the very human sound of vexation. I should stress here that it's not easy to hear. Some have transcribed the utterance as a simple 'oh no', but others hear the following: 'It's the eleventh day of March, eighteen hundred and eighty five. [Trilled R] How is this for high! Mary had a little lamb, and its fleece was white as snow, and everywhere that Mary went — oh, fuck.'[12] By March 1885 Tainter was being assisted by one Harry G. Rogers, and it is thought that Rogers can lay claim to this first recorded f-bomb.

Another disc, this time made of brass, has a recording of Hamlet's Soliloquy cut vertically into green-coloured wax, probably read by Chichester Bell. Again, from the surviving lab notes and diaries, we know they tried brass discs over several months, and that in one entry from March 1884, they decided to try adding some colour to the wax to enhance visibility. There are later discs of wax on cardboard, one of which includes a long description of the history of a cotton manufacturer. IRENE also managed to resurrect that sealed-box electrotype from 1881. The audible content consists of what is probably Tainter trilling some Rs and saying some numbers slowly and deliberately.

The Volta associates had taken new paths towards logical conclusions and technological limits. They had turned their minds to problems that would stymie the fledgling record business for generations to come. They had tried techniques that were decades ahead of their time, testing light, jets, compressed air and magnetism. They had tried to overcome needle-to-groove surface noise and variable speeds. But all these experiments had used up considerable time and money. Now, with Chichester Bell making noises about returning to

---

[12] Transcribed by Patrick Feaster.

Europe, and elements of Edison's patent coming to an end, they needed to bring their lengthy researches to some kind of commercial fruition.

In the early spring of 1885, they made a graphophone that created records on long, narrow strips of wax-coated paper. Tainter realised he needed to narrow the groove and in May began testing a machine with 9-inch long wax-coated paper cylinders, sporting about 120 grooves per inch. Two months later the cylinder had shrunk to 6 inches, now with 150 grooves per inch, and by July he added hearing tubes. Prototype working, they ordered the construction of six machines for testing. This started in New York but, with work progressing too slowly, Tainter brought them back in house and the first generation of machines were finally completed at the Volta Lab at the start of 1886.

They worked well. Indeed, they proved so satisfactory that the members of the Volta Laboratory Association immediately decided to form a joint stock company, the Volta Graphophone Company, which was incorporated in February. The new investors were keen to tap the business market, and so the focus for this new audio tool was dictation – it came with a treadle motor, speed regulator, stop/start keys, a brush to clean wax shavings, and a new reproducer.

It didn't look all that different from Edison's original. The most obvious difference was that the cylinder stayed in the same position, while the carving stylus moved along as it made impressions in the wax. However, it scored over the tin-foil predecessor in several important ways. Achieving a steady speed was a cinch thanks to the treadle mechanism, and the graphophone's loosely mounted, floating stylus was much better than Edison's rigid reproducer at following the grooves. Wax was quieter than tin foil, true, but once armed with listening tubes, the quality was markedly better. The records themselves could be quickly removed and were cheap to despatch.

Armed with a device that sounded better and was easier to use, at some point in the build-up to launch, the Volta team contacted the Edison Speaking Phonograph Company, offering to co-operate and share in this new talking machine. The upstarts must have been quite annoying to Edison. After all, his original patents had included wax, 'or yielding material', as a means to support metal foil that then received the indentation. He began to pull up the drawbridge.

The skirmishes that followed revolved around one very important point of semantics. Edison's patents had used the word 'indenting' – which meant embossing the material, without the removal of any part of it – whereas what the Volta/graphophone lot proposed was *incising*, cutting into the material, in this case waxed cardboard. This may sit at the heart of why it took so long for the Volta team to go public. The tech they had was all but decided back in 1881, yet they continued to experiment, possibly in the hope of finding virgin ground from which to operate. It may be that they found more confidence in their position once their lawyer, a man named Philip Mauro, had begun to argue that there was a clear difference between the words 'indenting' and 'incising'. He was showing them that they had a defendable position. Edison rejected their advances and immediately set about improving his own machine.

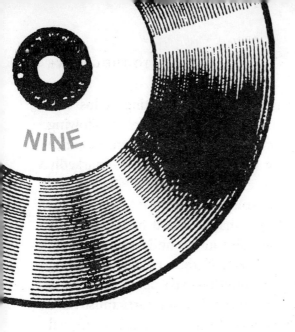

NINE

Edison Returns

'I now use a cylinder of wax for receiving the record of sound-pulsations ...'

Thomas Edison

The Graphophone failed to set the world alight. The machine was undoubtedly more usable than its predecessor, and it certainly won support, but it was still a tricky customer, and just not as useful as it sounded like it might be on paper. Nevertheless, the Volta Graphophone Company was now in charge of its various patents, and the American Graphophone Company had been formed to make and sell the machines. They looked set to dominate a brand-new industry.

Once the graphophone patents had come through, the Volta team began marketing in earnest, and there followed public demonstrations and lengthy illustrated articles in specialist press and national newspapers. A sewing-machine factory in Bridgeport, Connecticut, became American Graphophone's manufacturing plant, and Tainter lived and worked there for several months. To begin with they made about four machines a day.

Early graphophone advocates were reporters for the House of Representatives. This extremely high-pressure, time-critical job saw them attend live debates from the lower house, scribbling notes until they had enough material for a column in the next morning's edition of *The Congressional Record* – the official record of the proceedings, published daily when Congress is in session. The reporter would take his shorthand notes to a shorthand clerk, read them out, and the clerk would then take them down in more shorthand before finally transcribing them on a typewriter. Though complicated, tedious and time-consuming, this process did guard against errors. Once they got hold of a graphophone, however, it was possible for the original reporter to quickly read out his own notes into a dictating machine, then listen back and check it was all correct. Then he could hand the cylinder over to any old typist, doing away with the need for a specialist shorthand clerk.

The American Graphophone Company made their first official approach to Edison, demonstrating the improved graphophone to two of his representatives at the St James Hotel in New York. Tainter had worried this would spur Edison into action, and he was proved correct. Edison was by now set up in a new laboratory in West Orange, New Jersey, and he immediately diverted his energy towards improving the phonograph.

He would cultivate a narrative in which the genius creator had never entirely forgotten his beloved invention, tinkering with improvements in the intervening years that were now beginning to bear fruit. This wasn't the case – he had turned his back on the phonograph completely, allowing an important patent to lapse simply through not paying a $100 fee. But in the end, the graphophone was beaten back by a succession of new machines from the Edison stable – the New Phonograph, the Improved Phonograph and, finally, the Perfected Phonograph.

It wasn't all plain sailing. Edison gave a demonstration of one new prototype machine in the spring of 1888, seeking to impress a group of potential investors. He dictated a letter into his machine and set the reproducing stylus in place. Instead of hearing his words, the machine hissed. He started again, put in a fresh cylinder, dictated a second letter. After a second instalment of hissing, the investors decided to pass. Despite these setbacks, Edison finished his Perfected Phonograph after a 72-hour marathon beginning on 13 June and ending on 16 June. The aftermath was captured in the famous photo by W. K. L. Dickson, taken at 5.30 a.m., showing an exhausted, dishevelled Edison with his ear to a listening tube.

The new generation of machines were similar to the graphophone, using wax and a floating stylus, as well as battery-powered motors to achieve constant speed. Edison had employed Franz Schulze-Berge specifically to develop a wax in answer to the carnauba mixture that Tainter used. Instead of a layer of wax on cardboard, however, these Edisons were carved into solid wax cylinders. This meant that a cylinder could be shaved and reused over and over, which seemed to be a feature that would prove advantageous for office use. The electric motor, too, sounded good on paper, although it involved a lot of messing about with bichromate cells.

With the reproduction of accurate pitch now achievable, some of the earliest serious attempts to record music were made. Child prodigy Josef Hofmann, born the year before the tin-foil phonograph was invented, was the first pianist to record a cylinder, when he paid a visit to the Edison laboratory at the age of 12. Later in the same year another pianist, Hans von Bülow, recorded some Chopin on the newfangled apparatus. The story goes that, on hearing his work played back via the ear tubes, he fainted.

The Bell-Tainter machines were built upon a patent that 'engraved or incised' the recording medium, sidestepping infringement of Edison's original, which merely 'embossed or indented'. Edison and his advisers had concluded this argument, though very annoying, was strong. However, there was no gentleman's agreement here. When his competitors saw the sparkling-new Edison machines, they saw infringing devices that blatantly carved away the wax and used *their* floating stylus.

As patent litigation is very much a fun-free zone, it's good news for us that at this point a Pittsburgh businessman named Jesse Lippincott headed off a good deal of the trouble with some much-needed cash, bringing the enemies together ... for a bit. He invested $200,000 in the graphophone business, for which he bought himself exclusive rights to sell the machines in the United States.

And two weeks after being photographed looking tired, Edison signed a contract that sold his company to Jesse Lippincott for $500,000, forming the North American Phonograph Company. So by the end of 1888, Lippincott and his company were in complete charge of the fledgling talking machine industry in America. In short: Edison Phonograph Works had exclusive rights to manufacture phonographs and accessories, while North American had the exclusive rights to market them, via various regional sub-companies.

According to Roland Gelatt, author of *The Fabulous Phonograph* (1955), Lippincott himself was not the ideal man to be in charge. He had been attracted to the machines by their promise of use as an office aid. He dreamed of a future without stenographers and was convinced the business world would pay big bucks for this new utopia. Once on this path, he was blinkered to all other possible uses for the machines. He also, unwisely with hindsight, modelled his business on what was working for the telegraph and telephone industry. He divided up the country into territories, selling off the rights to trade in phonographs in these areas. He also chose to lease machines, rather than selling them outright.

The first year was pretty bad. The franchises were set up, yes, but all but one of the new subsidiary companies lost money. The new factories making Edisons and graphophones[13] both ran into production problems,

---

[13] The same summer the graphophone went on sale, Tainter was married. A bout of pneumonia, caught while overseeing the building of the graphophone factory in Bridgeport in 1888, would incapacitate him intermittently for the rest of his life. He worked

and the leased machines that *had* made it out frequently stopped working and had to be repaired, at great expense, or sent back, at great expense.

The machine just wasn't proving to be the world-changing device that people thought it might be, in the office at least. An expensive, clunky, unreliable and hard-to-use machine was more annoying than a freelance stenographer. The only company, out of that flurry of regional enterprises, to make a profit and pay a dividend was the Columbia Phonograph Company. And again, according to Gelatt, this was mainly because it had come about through an already established business that was servicing graphophones being operated in government offices – one of the few places where the machines had proved useful.

Lippincott bowed out[14] and Edison assumed charge. He abandoned the lease-only model and soon began selling machines outright, although he proved to be similarly blinkered when it came to the uses of his 'baby'.

From around 1889 it became increasingly obvious that whatever Lippincott, Edison, Bell or Tainter thought, there was a market for the phonograph as a

---

in New York City for a time, but became ill again as he was preparing for the Chicago World's Fair. He eventually enjoyed a long semi-retirement in San Diego.

[14] Following Lippincott's bankruptcy, the American Graphophone Company was saved by an alliance with the Columbia Phonograph Company. Columbia then became the sole worldwide distributor of the next generation of graphophones, intended for home use. Edison would eventually be forced to liquidate the North American Phonograph Company. During these bankruptcy proceedings, he was legally prevented from selling machines in the United States.

means of entertainment. Canny fairground operators were fitting phonographs with U-shaped hollow pipes connected to several listening tubes. They could then charge multiple punters to listen to a single cylinder. Coin-operated versions were even better. The first 'juke box'-type machine was installed at Palais Royal Saloon in San Francisco on 23 November 1889. It was invented by Louis Glass, another Western Union telegraph veteran, whose office was just two blocks away. Not much is known about that first machine other than it seems to have been very profitable – Glass claimed it earned more than $1,000 in six months (equivalent to approximately $32,561 in today's money at the time of writing). During a trade conference in Chicago in 1890, Glass told the assembled operators and manufacturers that his first 15 machines had brought in $4,000 from December 1889 to May 1890. This prompted hundreds more to start setting up nickel-in-the-slot phonographs, boasting returns of $50 to $100 a week, while one in a New Orleans drugstore claimed an average take of $500 a month. These were, at last, excellent returns.

For the next few years, Edison continued to bang on that his machine was a crew cut not a longhair. It wasn't a toy and it certainly wasn't a musical instrument. The first issue of in-company journal *The Phonogram* captures the Edison party line of the period: slot machines were a bad idea. Any partner that relied solely on coin-in-the-slot machines for its income was making a fatal mistake that would cost them in the end. It urged its readers to take every opportunity to push the machines towards business. And it warned that every time someone used a nickel phonograph, the more damaging this would be to its long-term reputation. The phonograph would eventually be laughed at, dismissed as a frivolous toy, not a serious tool.

Although Edison would maintain this anti-fun stance in public, under the new arrangement his phonograph works in Orange had to manufacture phonographs, accessories and 'special extras'. And these extras included recordings of music.

The 'First Book of Phonograph Records' is a handwritten logbook of the musical recording programme at the Edison laboratory that took place from 1889, and as such has become known as the 'founding document' of the recording industry. It was kept by a German named Adelbert Theodor Edward Wangemann – better known as Theo. He was about to cross the Atlantic to keep a phonograph on tour to the 1889 Paris Expo in working order. But in spring that year, having already been experimenting for months, he began logging recordings.

The first page, dated 24 May, lists a selection of tracks by 'Mr. F. Goede' on a flute, starting with the 'Famous 22nd Regiment March' by Patrick Gilmore. This is closely followed by 'The Warbler', the 'Lilliput Polka' and a 'Bird's Festival Waltz'. The next day they tried recording a violinist, the third recordee was a cornet soloist named John Mitthauer, on 29 May they tried a trio – clarinet, flute and bassoon – and in early June they recorded a flute trio.

One of the earliest surviving documents we have of the burgeoning record industry, outside this logbook, was a North American price list from May 1889. Aimed at North American's regional sub-companies, it lists prices for various supplies, including mixed 'musical phonograms', in boxes of six or twelve. That's it. There's no artist names, no song titles, not even a mention of the types of music or the instruments used. North American and the sub-companies complained to the Edison laboratory that this wasn't enough to go on. 'Give us a proper catalogue so we can choose what we want,' they said. Eventually, Edison's works obliged, and in mid-January of the following year North American published its *Catalogue of Musical Phonograms for the Phonograph*. However, just a week later, apparently sick of requests and complaints about the recordings, Edison announced that he was to discontinue this arm of their operation altogether. He was basically saying: 'Fine. If you don't like our clarinet, flute and bassoon trio, make your own then!' By February 1890, both the New York and Columbia companies had published their own versions of the catalogues, with additions.

This then is the means through which the modern recording industry was born – via side-show listening tubes, coin-slot jukeboxes, and grumbling directors who would really rather you didn't. With the nickels rolling in, companies continued to ignore grouchy grandad and instead look for new, popular content. Despite the limited tonal range, short length and labour-intensive production methods, the public wanted more.

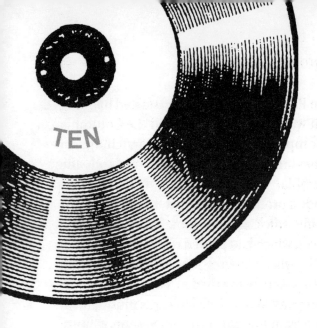

TEN

Performers and Producers

> 'Remember! "Casey" Records are loud, clear and distinct, and are especially recommended for horn use.'
>
> Russell Hunting, 1896

The Music industry of the 1890s had a rustic, DIY aesthetic. One of its early stars cut many of his most popular cylinders at home, accompanied by a schoolboy on piano. For a time, Washington, DC, New York, Chicago and Philadelphia were the global epicentres of this new cottage industry, and the world's first recording stars were a brass band and a whistling government clerk.

Edward Easton was a court reporter-cum-lawyer whose work meant he was naturally attracted to the labour-saving promise of the graphophone. Once the

North American Phonograph Company started handing out franchises, it was Easton who founded the Columbia Phonograph Company, which put him in charge of talking machines in Maryland, Delaware and the District of Columbia.

Columbia made a profit from the start, but only just. At first the company stuck to the game plan of leasing machines to offices, which had proved relatively popular in Washington, DC – when Congress was in session between 50 and 60 machines were being used in The Capitol alone. However, the company soon pivoted towards music and in particular income from the coin-in-the-slot phonographs. By May 1891 they were running 126 of them.

Columbia would eventually move from temporary offices to a handsome five-storey brownstone building at 627 E Street, North West, Washington, DC. The first floor housed Easton and his managers, the repair shop was in the basement, the second floor had displays and coin-slot machines and the third floor was the music department. An army of inspectors on bicycles would be sent out to check on commercial and automatic phonographs dotted throughout the city. And by the time the May issue of *The Phonogram* was released, the company was working with Dr Richard Rosenthal, overseeing the development of language-teaching cylinders. (He had opened a school of languages nearby, where pupils could be seen studying with books in hand, tubes in ear.)

The new generation of phonographs – louder, easier to use, with better sound – were appearing in social clubs, hotels, funfairs, theatres and public spaces. But from the mid-1890s onwards, more models were being aimed at the domestic market, as the middle and upper classes rushed to acquire their own talking machines.

The spring-motor powered Baby Grand Graphophone of 1896, for example, was a package designed for 'exhibition or home entertainment'. This was an expensive item, but for $100 you got a graphophone, nine records, three blanks, three hearing tubes, a large reproducing horn and a canvas case for holding the records.

Making records was long, laborious and hot. Every one was unique, an original, meaning that if you wanted to sell 100 records, the performer had to perform 100 times. From the viewpoint of being a customer, of course, this meant a unique intimacy between artist and listener – your own personalised take. The only way around this problem was to sing into more than one recording machine at the same time, which is exactly what they began to do. Some purists in that first generation of sound engineers could be rather sniffy about the practice. To obtain good recordings, they argued, you needed to arrange musicians and singers carefully, and producing multiple copies was contrary to the art of phonography.

In the pre-microphone studio, performances were generally loud, or clear, rather than of great artistic merit. To reproduce well on a cylinder, they needed to blast out the favourites. There was very little intimacy. Singers and producers learnt to adapt themselves to the new medium – sticking their heads practically inside the horn for the quiet bits, withdrawing for the loud bits. Musicians, meanwhile, would be tightly packed around recording cones, frequently clashing elbows, bows and bells.

Many singers achieved success not because they were the leading performers of the day but because they had stamina and a voice that translated well to wax. Antonio Pozo, a local sensation in Madrid, Spain, around the turn of the century, made hundreds of recordings at studios across the city. After one persuaded him to sign an exclusive contract, a rival studio immediately had him record multiple copies of all his greatest hits in one marathon session that lasted from 7 p.m. to 7 a.m.

While renowned artists were persuaded to record, many simply didn't like it or refused outright. Some hated how they sounded on record. Others disliked the feeling of being recorded, which felt strange, unfamiliar and challenging even to those who had spent a career on stage. A report from a New York studio included an interview with the engineer: 'It is often difficult to get the proper attitude on the part of the singer ... There is such a thing as stage fright in performing for the phonograph. I do not know how to explain it, whether it comes from the thought that the record will be reproduced far away from the singer's presence and perhaps long after he is dead or from some other reason.'

It was all rather labour intensive, too. And if you're being paid top dollar to tread the boards in a plush theatre in front of a receptive audience for a couple of hours in the evening, the idea of having to perform the same song over and over again, in some small, smoky studio on a rainy Tuesday afternoon, can't have been that attractive. This is in part why, in those early days, the first stars were often working singers and jobbing musicians without any kind of international reputation, who were simply trying to earn extra money.

Cylinders didn't work with all forms of music. Certain instruments didn't translate well either. Loud, percussive brass bands were in, and whistlers were popular, as were clearly enunciated comic monologues. In the early days engineers found it harder to record female singers with any success. 'Their high and fine tones are apt to shrill and shatter when transferred to the rolls,' explained one correspondent. 'That difficulty has been overcome now so that many women as well as men are recorded. But it remains a fact that some singers, men as well as women, never give us very satisfactory results. It comes, I suppose, from some peculiarity in their mode of vocal expression.'

The United States Marine Band, known as 'The President's Own' ever since performing at the inauguration of President Thomas Jefferson on 4 March 1801, was based conveniently close to the offices of Columbia. John Philip Sousa was a boy with perfect pitch and a violin when he was first enlisted as an apprentice in the Marine Band after attempting to join a circus. He served for several years while also studying music theory and composition. During the 1870s, he left to play with and conduct various touring and theatre orchestras, before returning to lead the Marine Band in 1880, at the same time composing some of its most loved marches.

Under Sousa,[15] the Marine Band made its debut recordings with Columbia. The first Columbia record catalogue, from 1890, is a one-page listing of 'superior in loudness, clearness and character of selections to any

---

[15] Sousa himself was such a draw that a promoter, David Blakely, convinced him to resign and form a civilian concert band. It was known as Sousa's New Marine Band until several complaints from Washington persuaded him to drop the 'Marine' reference. In 1896, while on holiday in Europe, Sousa got word that his promoter had died. And on the return voyage he was inspired to write his most famous composition, 'The Stars and Stripes Forever'.

band records yet offered', including several by its star attraction. Soon the band was making records for Columbia on the third floor in Washington, DC, performing in front of banks of 10 or more recording machines. (It's interesting to think that, while the resulting cylinders would have been sold as identical, in fact each would have its own unique quality because of the position and angle of the different recording horns.) The band cut 60 cylinders in the autumn of 1890 – and remember, each one of those would have represented multiple performances – and by 1897, more than 400 Marine Band titles were available for sale.

Such was the popularity of military pomp that each of the regional companies soon had their own brass bands. New York's home-grown phonographs of the early 1890s generally came through the hands of Charles Marshall, a tidy, obsessive character with a neat side-parting, wraparound moustache and sideburns. He had spent the winter of 1889 recording church chimes, installing two machines in the steeple of New York's Trinity Church, at the intersection of Wall Street and Broadway. By the end of the season he'd made 1,200 records. His most popular export, however, was Capp's Seventh Regiment Band of New York.

Until well into the 1900s, commercial recordings would typically contain a spoken announcement of what was to follow. Marshall believed that a clear and distinct introduction was key to selling records. Musical records are 'half made by a perfect announcement ... Nothing is more gratifying to a listener,' he said.

A typical brass band session at the New York Phonograph Company, 257 Fifth Avenue, is described in an issue of *The Phonogram* from March 1891. In one room, the band is surrounded by 10 phonographs, each with a large recording horn. An attendant has already checked all the cylinders and the batteries are loaded and working. Marshall stands before the first machine and records a good, clear introduction, naming the tune and the band. Once complete, he stops the recording, steps to the second phonograph, sets it recording again, and repeats his introduction. Having completed 10 introductions, all the machines are started simultaneously, and the band begins to play. A close eye is kept on the cylinders as they go. If they are running short of space before reaching the end of the tune, Marshall simply raises a finger, and the music stops at the end of the next phrase. If they complete a tune, they sometimes fill dead space with cheering and applause. By the end of a three-hour session, all being well, they might have 300 cylinders ready for sale.

Early phonography, then, was a strange melting pot of whatever worked, and the average studio was something like a cross between the backstage of a theatre and a machine shop. The gelatine-like blanks would arrive in long wooden boxes, placed on trays above steam pipes to keep them warm and ready for sound. They would record laughter, trains, jokes, chimes, whistling, skits and sketches. There were a lot of what writer Clive Thompson describes as 'acoustic blackface', songs where white performers would imitate African Americans, such as Arthur Collins' 'The Preacher and the Bear', in which a terrified preacher is chased up a tree, and vaudeville veteran Billy Golden's hit 'Turkey in the Straw'. 'Uncle Josh Weathersby's Visit to New York' had a country 'hick' visiting the big city, and these proved so popular that its creator, Cal Stewart, spent more than two decades recording more adventures of his beloved rube, beginning each one with a signature burst of laughter. Companies were also experimenting with a kind of 'audio theatre', where skits combined multiple speakers, effects and mimicry to create an atmosphere. The later Edison Standard release, 'Christmas morning at Clancy's', for example, starts with the usual announcement before a festive combination of church bells, door knocking and children calling 'Merry Christmas' takes place.

George W. Johnson, the son of a freed slave, was making money whistling popular tunes on street corners and ferry terminals in New York, where he was spotted and recruited by distributors Charles Marshall (of the New York Phonograph Company) and Victor Emerson (of the New Jersey Phonograph Company). His first recordings for both were the popular vaudeville novelty song 'The Whistling Coon' and 'The Laughing Song'. By 1895 these had become America's biggest-selling records.

One early hit, 'The Mocking Bird', begins with whistles, before John Yorke Atlee barks: 'The Mocking Bird! Whistled by Mr John York Atlee! On this equipment for the Columbia Phonograph Company of Washington, DC, accompanied on the piano by Professor Gaisberg'. This is notable for containing two important figures of early sound. John York Atlee was a government clerk by day, a deep-voiced whistler by night. 'Professor Gaisberg' was a 16-year-old named Fred Gaisberg, a character who accompanied many of the popular recording stars of the day, bore witness to the first generation of commercially successful cylinders, then crossed the tracks to become an influential producer and talent scout.

Looking back, Gaisberg laughed at Atlee's pompous announcements, delivered in tones that might make those listening imagine a giant when in reality he was a 'shrimp of a man', a 5ft clerk who was just a little too proud of his fine flowing moustache. Atlee's office hours were nine till four. After work he would return home, where he was joined by the just-about-out-of-school-age Gaisberg. In his parlour was an old upright piano and a row of three phonographs lent by Columbia. There, in batches of three, they would belt out 'The Mocking Bird' or 'The Laughing Song', and many others. If Gaisberg was busy, Atlee would simply make solo recordings such as readings of Mark Antony's speech from Shakespeare's *Julius Caesar* and the Lord's Prayer.

Fred was born in Washington, DC, in 1873. His father, a German immigrant who had arrived in the United States in the 1850s, worked as a bookbinder. By the time Fred was 10, the family had moved to a house on New York Avenue and he began learning to play on his mother's upright piano. As a youngster he was a chorister at St John's Episcopal Church and also performed in Sousa's choir, put together for Sunday evening concerts, attending rehearsals at Sousa's house in the Navy Yard in South Washington. As a schoolboy, his skill on the piano won him a scholarship and soon led to regular gigs at amateur and charity events. Then, in the summer of 1889, he heard tell of a novel way to make a little extra money. Columbia were sick of not making much replacing stenographers and instead wanted to make some music for coin-slot machines – they needed players, and specifically someone who could play a piano loudly and clearly.

After graduation, Gaisberg asked Columbia for a full-time job. They weren't at all sure that they needed the headache of actually employing a full-time musician, but they came up with a compromise: yes, he could join the company, but only if he learnt how to produce as well as play. So he was sent on a kind of extended technical boot camp, working at the North American Phonograph Company's factory in Bridgeport, Connecticut, where, under the eye of Scottish engineer Thomas Hood Macdonald, he was introduced to cutting-edge phonography. Here he learnt many of the technicalities of recording and witnessed some of the latest gizmos to help keep cylinders revolving at a constant speed.

Fred was soon back in the capital, continuing to cut records for coin-slot machines, recording endless takes of Atlee favourites, as well as working alongside other popular performers of the Washington scene, including a bass-voiced Irish man named Dan Donovan, who during the day announced train departures at Potomac railway station and at night recorded the likes of 'Rocked in the Cradle of the Deep' and 'Anchored'.

Through Donovan, Gaisberg was introduced to Tainter, who at that time was working on a new machine that could cut up to 20 cylinders from a single performance. Fred was fascinated by this precise, delicate man, who between cups of tea would forever experiment, making tiny adjustments to diaphragm or cutting stylus in his search for perfection, stressing sibilants into the mouthpiece, as S-sounds were the hardest to pick up. If any record bore the slightest indication of an 'S', he would smile to himself and have another cup of tea.

Tainter had just signed an agreement to provide coin-operated machines for the Chicago World's Fair, and so he needed content. Earning $10 a week, Gaisberg began harvesting local talent. He would load up Tainter's bank of 20 machines with wax-coated cylinders, arrange the wall of recording horns, then accompany artists on the piano. The repertoire was dominated by recent hit 'Daisy Bell' and 'After the Ball Was Over', and sometimes they performed the latter as many as 70 times a day.

Though they got the coin-slot phonographs filled and installed at the World's Fair in time for its grand opening in May 1893, the machines were not a success – they were just too unreliable – and so they were shipped back to Washington, where instead Gaisberg was for a time in charge of installing them in saloons, restaurants and beer gardens. Their unreliability continued to be a pain, however, and they would often accept coins without playing a tune. Gaisberg, who had to reload and collect, sometimes bore the brunt of disgruntled bartenders.

While his time with Tainter had in some ways been a failure, it made Gaisberg recognise the value of his experience and contacts. He had been able to muster a collection of recording talent. Fred returned to Columbia, who by now had moved into their five-storey home on Pennsylvania Avenue, which had an Audition Saloon, about a hundred automatic nickel-in-the-slot phonographs with rubber listening tubes, and a top-floor recording studio. Here he accompanied handsome baritone Len Spencer, Billy Golden, George Gaskin,

Daniel Quinn, George W. Johnson and Johnny Meyers, as well as overseeing recordings by the Marine Band and the Boston Cadet Band.

Another star, and arguably the most popular of all the pre-mass production artists in America, was Russell Hunting, whose rapid-fire skits involving the Irish character Michael Casey were so popular they sparked an entire army of imitators. Hunting was an actor with the Boston Theatre Company who had leased a phonograph for his own use, just to experiment. Certain types of voice might come out as garbled scratchings, but while experimenting he realised he sounded particularly good on record.

His debut record, with the New England Phonograph Company, came in 1891. Soon, he was recording for Columbia and many others. Versions of his 1892 skits 'Casey at the Telephone' and 'Casey Taking the Census' became his most famous works, as did the baseball poem 'Casey at the Bat'. Casey was so popular that Hunting lost control as more and more recording artists imitated Casey themselves, in poor knock-offs of the original. In a bid to stamp out this annoying practice, he launched his own magazine, *The Phonoscope*, in 1896. The first issue covered 'Voices of the Dead – the Possibilities of the Talking Machine', a report on the ongoing dispute between the phonograph and graphophone, and, on the front page, a vigorous defence of his territory.

> I do not copy, imitate or mimic others. I originate, manufacture and sell my own records. Remember! I am the originator of the Casey series. Remember! That there are 'Casey' Records on the market which are not

manufactured by me, but are made by others, using my subjects, in order to deceive the public. Remember! ... That certain unprincipled individuals and corporations are duplicating my work, thereby deceiving the public by furnishing a record about one-third as loud as the original. Remember! 'Casey's are the Standard Humorous Talking Records. There have been over 50,000 manufactured and shipped to all parts of the English-speaking world, giving universal satisfaction.

Although he didn't mention them in *The Phonoscope*, Hunting also put out a series of indecent recordings for saloons and amusement arcades on Coney Island and was briefly imprisoned for violating obscenity laws. He wasn't alone either, and lots of cylinder nasties appeared in the 1890s, until legislation caught up and outlawed them. While most of these lewd recordings were destroyed, you can listen to examples that survived in the 2007 compilation *Actionable Offenses: Indecent Phonograph Recordings from the 1890s*. I've listened to them, so you don't have to – it's mainly profanity-laden skits and limericks.

Sustained by its musical recording business, Columbia survived an economic depression in the mid-1890s, which sank many of the other local distributors. By pushing music cylinders to small exhibitors and the coin-slot trade, it was able to absorb one of the principal phonograph makers in the American Graphophone Company. In 1895, no longer hamstrung by being a mere regional operative, Columbia opened its first office and studio outside Washington, DC, in New York City. And in another example of weird skulduggery that only makes sense when you see the checkmate move, Edward Easton, now general manager of the American Graphophone Company, filed a lawsuit against his own company, Columbia, for selling Edison phonographs that his lawyers argued infringed on Volta's wax cylinder patent. The courts ruled against Columbia, which validated the original graphophone/Volta patent over Edison's improved phonograph. This forced Edison to sign an agreement with Easton in 1896 to share patents. This meant that, from then on, Easton could make and sell music cylinders of the Edison type, but under the graphophone badge. So it was that, at this point, the original treadle graphophone, with its narrow, 6-inch long cardboard cylinder, was discontinued as a means of entertainment.

The dictaphone that helped the phonograph over the line was dead, and the horizon was clear. Columbia and Edison could both make Edison-type cylinders, designed to be played on increasingly low-cost spring-motor machines – such as the $10 Eagle model that arrived in 1897. At that price, even with the shaky economic climate, lots of Americans could afford phonographs and records. Columbia's one-page catalogue from a few years before had grown into a volume, listing hundreds of titles in multiple categories, alongside pictures of artists and advertisements. The company had swallowed many of its old regional competitors and by 1900 had branches in Buffalo, Chicago, Philadelphia, San Francisco, St Louis, Paris and London.

ELEVEN

Crossing Continents

> 'It is now nearly one o'clock, and he and Van Helsing are sitting with her. I am to relive them in a quarter of an hour, and I am entering this on Lucy's phonograph.'
>
> 'Dr Seward's Diary', *Dracula*, Bram Stoker, 1897

In August 1888, the *South Wales Daily* heralded news that Mr Edison's phonograph would soon be rendered 'loud-speaking'. 'Capable, that is to say, of communicating its message to several hearers at once without the intervention of tubes extending from the instrument to their ear.' It reports from a live demonstration given by one Colonel Gouraud to a party of visitors, exhibiting a second instrument that he had just received from

America. This phonograph, 'being furnished with a sort of speaking-trumpet, from which the sounds issue, is distinctly audible to a large group of persons.' The reporter describes some of the sounds played, which include 'habitual noises' from Edison's US workshops, including the sounds of anvil, sandpaper and a winding telegraph, then a snatch of speech from *Faust*, before a final celebrity cameo in the form of virtuoso whistler and touring sensation Alice Shaw, known as 'La Belle Siffleuse', the phonograph reproducing her notes 'with astonishing accuracy'. Colonel Gouraud then informed the gathered guests and pressmen that he was in the custom of dictating letters direct to the phonograph at his leisure, later setting the instrument in motion and writing from its repetition. 'The actual presence of a shorthand clerk is thus no longer necessary ...' finished the report excitedly.

Phonographs were spreading all over the world through legal and illegal channels. Gouraud was a Civil War veteran who had won the Medal of Honor at the Battle of Honey Hill. The colonel had already proven himself as an able promoter of Edison's products, not least the telephone and the electric light.[16] He had installed himself as the inventor's London mouthpiece, christening his Norwood home 'Little Menlo', which also became known as 'the Electric House' among locals, as it was full of cutting-edge gadgets. Doors opening

---

[16] His association with Edison was in a sense a return to his family roots – his father, François Gouraud, had crossed the Atlantic several decades before to spread word of the new daguerreotype technology.

would trigger electric lights, his carpets and boots were cleaned using electrical devices, he travelled the capital on an electric launch and a battery-powered tricycle, and his billiard room had been converted into Britain's first proto-recording studio.

Gouraud hadn't been a passive presence when it came to the phonograph. He had been piling pressure on Edison to get the new phonograph completed and sent to him. When the Edison Perfected model finally arrived in 1888, the colonel took to it with gusto, gathering prominent names of the day to show it off, recording all manner of spoken word, performed music and public events, setting a number of world-firsts in the field, generating a flood of column inches, and announcing a project to compile what he called a 'Phonogramic Album' of famous voices of the day. And when he formally introduced the new Edison Phonograph to London's assembled press, he played one of the earliest recordings of music to survive – a piano and cornet duet of Arthur Sullivan's 'The Lost Chord'.

One particularly important event was the 'phonograph party'. This is remembered not only because it was widely reported at the time, but also because the audio has survived and, thanks to a forensic analysis carried out by early sound scholar Patrick Feaster, we now have a blow-by-blow account of what happened.

It was a strange, highly choreographed event that took place amid wines, meats and cigars at Little Menlo. Guests of honour that night included Arthur Sullivan (this was just days after the debut of the opera *The Yeoman of the Guard*), Edmund Hodgson Yates (editor of upper-class society paper *The World*) and dramatist Augustus Glossop

Harris (known as 'Druriolanus'). The entire event was built around the phonograph, which, by use of a carefully prepared series of pre-recorded cylinders, the colonel was making both toastmaster and speechmaker.

The guests were already fed and watered when the machine was dragged in by the colonel's helper, a socialite named Alexander Meyrick Broadley. It was positioned on a table behind the host, powered by a Schanschieff primary battery. The phonograph, in fact the recorded voice of Broadley, would begin by calling for silence, then a toast would be introduced. Each toast had a theme and would start with some sort of preamble by the colonel, before a call to fill glasses and 'bumpers', ahead of the climactic cry 'to Literature!' or 'to Music!' This toast would be heard, along with the name of the guest who represented that particular world. Finally, the pre-recorded toastings would end with a flurry of cheers.

Following these machine-made toasts, the guests were, at some point in proceedings, played a cylinder with Edison's own voice before each was shuttled off to another room to record their own message for Edison. Yates managed to say: 'If I lack words to describe the dinner, it is because I am so enrapt and so enchanted by your invention that I find myself much more stupid than I ought to be after the grand excitement of our friend's meats and wine.' Then Arthur Sullivan spoke: 'I can only say that I am astonished and somewhat terrified at the result of this evening's experiments: astonished at the wonderful power you have developed, and terrified at the thought that so much hideous and bad music may be put on record forever.'

Another magical Gouraud recording was captured on a summer's afternoon in 1888. A performance of Handel's oratorio *Israel in Egypt* was taking place in the now-vanished Crystal Palace in London. The colonel was there, stationed in the press gallery, with his phonograph and a batch of the yellow paraffin cylinders. This represents one of the earliest deliberate recordings of a live musical performance known to exist – some earlier recordings from the 1870s are considered lost – and thankfully, the audio can be reconstructed with some degree of certainty as sound historians know the pitch of the Crystal Palace organ at the time.

The sound quality, by modern standards, is dreadful. It sounds like a microphone has been placed under a mattress and then submerged in a pool of standing water. But just as the power of the first photograph, taken out of an upstairs window in Burgundy in 1826, is only accentuated by its grainy, dream-like quality, so this snatch of ghostly lo-fi is like putting your ear to the door of 1888. It sounds close, yet far off. You can imagine you're running late to attend the performance, and you're hearing the far-off sounds of 4,000 voices rising in song, accompanied by a 500-piece orchestra as you approach the venue, impatient to take your seat. This is why I strongly suggest you take a pause from reading, and track down the recording. (Search: 'Crystal Palace 1888 Handel.')

Over the coming months, Gouraud invited distinguished personages to record their voices at Little Menlo. Henry Irving visited in August, and Gouraud described the occasion a few months later, remarking how the actor had stepped resolutely towards the machine, brimming with confidence, only to be hit with a kind of stage fright when faced with posterity. 'It is curious to see how the most distinguished speakers ... behave when they find themselves in front of the phonograph.' Irving too wrote an enthusiastic account, in a letter to his leading lady Ellen Terry: 'You speak into it and everything is recorded, voice, tone, intonation, everything. You turn a little wheel, and forth it comes, and can be repeated ten thousand times.'[17]

Gouraud had by now assembled a team of recordists who travelled the country, promoting the device and capturing the voices of celebrities. A scandal had erupted about the poverty endured by some Crimean veterans and, as part of the drive to raise funds for these destitute ex-soldiers, the colonel dispatched one Charles Steytler to Freshwater on the Isle of Wight to record Lord Tennyson reading 'The Charge of the Light Brigade'. Another captured the mournful playing of Martin Lanfried, the man who sounded the bugle at the Battle of Balaclava. Another visited 10 South Street, Park Lane, London, to record Florence Nightingale, who said: 'When I am no longer even a memory, just a name, I hope my voice may perpetuate the great work

---

[17] Irving's recordings were preserved in a box of white wax cylinders that were eventually acquired by the BBC in 1951. These were thought to have been looted from a bombed house in a street near luxury department store Harrods during the Second World War. They reveal that he chose to record *The Feast of Belshazzar* by Sir Edwin Arnold and the first stanza of *The Maniac* by Matthew Lewis.

of my life. God bless my dear old comrades of Balaclava and bring them safe to shore. Florence Nightingale.'[18]

The Edison Perfected phonograph was louder, more reliable and easier to operate. It was loud enough that it could be installed at inns, hotels, funfairs, theatres and other public spaces (although this didn't really get going until the 1890s). Bought or leased models would be paraded around by exhibitors, and more and more of the middle and upper classes were rushing out to acquire their own.

During this period the Edison and Tainter-Bell patents were combined in one iron-handed grip. While the colonel was peddling abroad with Edison's say-so, there were others who weren't sticking to the rules. Percy Willis, who would eventually work at the Edison Bell Works in Peckham, London, did not transport the first phonograph across the Atlantic. He was quietly confident, however, that he was the first to smuggle one across the Atlantic.

Willis and his partner were working in Canada, and not doing very well. Walking the streets of Montreal, they chanced upon a crowd hovering around a kind of curtained booth, outside of which was a poster announcing that a phonograph could be heard at a few cents a head. They took a train to Boston the very next

---

[18] This recording too remained 'lost' in a mahogany box in the Edison Bell archives for many years. It was discovered in the 1930s, eventually making its way to the Wellcome Trust collection. It was digitised in 2004.

day, securing one machine and three dozen cylinders, and signing all manner of 'portentous looking documents', binding them to all kinds of restrictions. The most important of these was that neither the machine nor the records were to be taken out of the United States. With the last of what cash they had left, they paid for passage from Boston to Queenstown, Ireland (the town was renamed Cobh during the Irish War of Independence).

Things didn't go well to begin with. On disembarking in County Cork, a strap supporting one of the parcels of records broke and 12 discs were destroyed without ever being played. Down to their last cash, they had no choice but to set up shop straight away. They still had 24 records left, and so installed the battery-powered machine right then and there, and people were soon 'rolling in by dozens and scores'. To begin with, armed only with a single listening tube, they had great difficulty in subduing the impatience of the crowd. Indeed, such was their sensational arrival in Queenstown that their fame preceded them as they travelled on to Cork, then Waterford and Limerick. When they finally opened at the Central Hall in Dublin, they took £200 in five days.

Willis and his partner chanced another smuggling expedition to the United States, bringing back more records and machines, this time hidden in apple barrels. They sold their contraband machines at inflated prices, generally to other travelling showmen. They toured with their own machine to Liverpool, to London, to the Isle of Man, interspersed with more trips to America, each one more profitable than the last. Music hall star Charles Coborn was enjoying great popularity at the time with the comic song 'The Man Who Broke the

Bank at Monte Carlo'. When Willis managed to get a record of the song, it became one of their most profitable cylinders. 'People used to come again and again to listen to it, and then they would bring their friends. Everybody wanted to hear that song, till at last the cylinder got clean worn out.'

In Germany, the fledgling trade followed a similar trajectory to that of the United States, France and Britain. Theo Wangemann was Germany's answer to Colonel Gouraud, demonstrating Edison's phonograph to Kaiser Wilhelm II in 1889. One early phonograph maker, A. Koltzow, focused his efforts on marketing the phonograph as a labour-saving device, before switching lanes, creating a design specifically tailored for the travelling showman. However, Germany's great gift to Europe in the early days of the phonograph was the 'Puck'. This was a simple, often exasperating but also inexpensive machine that truly opened up the world of phonographs to the masses. F. Ruppel & Co produced the Puck phonographs from the 1890s to around 1906. Pucks were solid and sat on cast-iron bases in the shape of lyres. Each had a wind-up single-spring motor with a one-piece starting lever and speed control. The first Pucks were designed with sound reproduction in mind, rather than as a recording device, which made things a lot simpler. The proud owner slotted the cylinder into place, started it rolling, then placed the all-in-one horn and stylus into position, the sounding end of the horn on a little stand, the stylus end, obviously, direct on the rotating surface. However, this meant that the unwary Puck DJ would know that a track had finished by the sound of the horn falling off the end.

Pucks sold in hundreds of thousands. In Britain, early models were priced as low as 3s 6d[19] and were even given away with qualifying purchases of cylinders. At such low prices, they made little or no profit, but they gave listeners the bug, and dealers the promise of future sales of either the cylinders to play or the more expensive players. The more precarious patent situation in the United States meant the plucky Puck didn't make as much of an impression over there. But in Europe the machine evolved and grew, adding spirit levels, flower-shaped horns, the ability to play later Grand-sized cylinders, feedscrew models for recording, a model that could play both cylinders and discs, and the beautiful

---

[19] To those unfamiliar with pre-metric Britain, values less than a pound were expressed in shillings and pence. A shilling (also known as a 'bob') was worth 12 pence. So if something cost 2/6 or 2s/6d, this indicated 'two and six', or more completely 'two shillings and sixpence' – which in total made (2 x 12 + 6 = 30) 30 pence. Anything costing less than a shilling would be described in pence. If something cost seven pence, for example, it would be written as 7d. Why the confusing 'd'? This comes from the Latin word *denarius*. Twenty shillings made a pound sterling, meaning there were in fact 240 old pence in one pound. You may also come across references to 'half crowns' and 'ten bob notes'. A half-crown coin was equivalent to two shillings and sixpence – one-eighth of a pound, 30 old pence, or 12.5 new pence. A '10-bob note' was a 10-shilling note, worth (10 x 12d) 120 old pence. As 120 old pence is exactly half a pound, a 10-bob note is equivalent to a modern 50p coin. At this point I might cite the example of John Tenniel's illustration of *The Mad Hatter*, which bore the sign 'In This Style 10/6', meaning the hat would have set you back 10 shillings and sixpence, or 126 old pence. However, as I'm beginning to feel breathless and dehydrated, I'm going to leave it there.

'Lorelei', with a decorative copper-plated, antiqued base and bright green horn.

These were not the only low-end, low-cost products available. The Edison Gem, introduced in 1899, was a 'bargain phonograph' initially priced at $7.50. The earliest were black, had no case, were wound with a key and could play two-minute cylinders – although sound reproduction was spoilt by the noisy motor. Pathé too had their Coquet phonograph from 1903, which was promoted as a mass-market and low-cost phonograph but in fact was out of reach to the average worker at 35 francs.

Even machines at the other end of the scale were finding a home. The extremely large, extremely heavy and extremely expensive Graphophone Grand was brought to market just in time for Christmas 1898. It cost a staggering $300 and was designed to play giant 'Grand' cylinders of 5-inch diameter (rather than the conventional 2-inch), costing an equally eye-watering $5 each. This is what sparked Edison to create his own high-end offering, the 'Concert' Phonograph, in February 1899 for $125, which played the Concert-sized cylinders. The final word went to the monstrous Multiplex Graphophone Grand, a steampunk nightmare with a bulky triplet of protruding horns. For $1,000, the buyer had three recorders, three reproducers, a triple horn stand, three 36-inch brass horns, 12 grand records and six grand blanks. In sensationalist style the *Ford Wayne Sentinel* reported on a Persian minister who had (apparently) ordered one from the Shah of Persia in March 1901. 'Never, in the history of present making, has so original and elaborate a gift been made by a subject to his sovereign.'

Phonographs were reaching countries and cultures all over the world. A single issue of *Edison Phonograph Monthly* reported on a huge press event in Berlin, on 72 new records for the Mexican catalogue, on an Australian customer who recently purchased five Standard Phonographs (one each for his three houses, and two for friends), and a missionary who described taking his phonograph to Japanese soldiers and Russian prisoners in Hiroshima and Matsuyama. He played a cylinder of a Russian hymn, recorded by the Edison Grand Concert Band, and 'every man was on his feet', wreathed in smiles and choking back tears.

Posterity was a constant theme in the early history of the phonograph. For some, the coming of the more usable second- and third-generation phonographs inspired an instinct to record and preserve fading languages and music. The Library of Congress holds many hundreds of wax cylinders of songs and stories relating to and documenting Native American cultures, many in languages that are no longer spoken. Among these are the earliest known ethnographic field recordings, documenting Passamaquoddy songs and narratives recorded by Jesse Walter Fewkes in Calais, Maine, in March 1890. Another collection was made by the first professional Native American ethnologist, Francis La Flesche, who worked with Alice Cunningham Fletcher on recording Omaha and Osage cultures between 1890 and 1910. The British anthropologist Alfred Cort Haddon mounted an expedition to Torres Strait and New Guinea in 1898, resulting in 101 brown wax cylinders recorded in the Torres Strait Islands in Australia, and a further 39 recorded in what is today

Papua New Guinea, which survive at the British Library. These include secular and ceremonial songs – music performed during periods of mourning and funeral preparations – as well as 'spinning top songs'. Kolap-spinning had 'recently been the fashionable excuse for an island gathering', and various songs were performed while sitting in a circle and spinning the tops. And during the First World War, a German linguist named Wilhelm Doegen visited prisoner of war camps across war-torn Europe, recording languages, music and songs. Because of his incredibly detailed notes, we know that on 27 September 1917, at 10.25 a.m., a Belfast-born private in the Royal Irish Fusiliers named John McCrory began singing 'The Pride of Liscarroll'.

Folklorist John Lomax had always been fascinated by the songs he heard while working on his father's farm. Around 1907, while attending Harvard, this passion for cowboy songs was encouraged for the first time. He was studying under English professor George Lyman Kittredge, who told him, 'Preserve the words and music. That's your job.' This would ultimately lead to his 1910 printed anthology *Cowboy Songs and Other Frontier Ballads*, which included songs such as 'Jesse James', 'Sweet Betsy From Pike'[20] and 'Git Along Little Dogies'. During this time he also began recording music using an Edison phonograph, one early example being 'Home on the Range', sung by a black saloon keeper in San Antonio, Texas.

---

[20] Strother Martin sings this song during the 'going straight' sequence of *Butch Cassidy and the Sundance Kid* (1969).

Through a grant from the American Council of Learned Societies, Lomax would eventually set out in June 1933 on the first of his famous recording expeditions, along with his then 18-year-old son Alan. By that time they had acquired a state-of-the-art, 143kg (315lb) phonograph uncoated-aluminium disc recorder, which was installed in the trunk of their Ford sedan. This was a huge step up from the imperfect hand-cranked Edison cylinder John had used in the past, or the wind-up office Ediphone, which they had used to record voices but always resulted in a very thin, faint sound. They visited the Louisiana State Penitentiary at Angola, recording a 12-string guitar player by the name of Huddie Ledbetter, better known as 'Lead Belly', and during the next year and a half, father and son continued to make disc recordings of musicians throughout the South.[21]

---

[21] Digitisation and preservation initiatives mean that we can go digging around in thousands of hours of these very early field recordings right now. The Lomax Digital Archive (archive.culturalequity.org) has recordings and audio that you can explore by place, instrument or culture. A team at Indiana University has been digitally preserving a collection of 7,000 wax cylinder recordings, the majority field-recordings cylinders made between 1893 and 1938 from 60 different countries, as well as early home recordings, offering snatches of audio from domestic life more than a century old. I particularly recommend you explore the UCSB Cylinder Audio Archive (cylinders.library.ucsb.edu), which has an amazing series of curated playlists, covering ethnic and foreign recordings, speeches and readings, plus popular songs, vaudeville acts and comedic monologues. This includes early cylinder recordings from Mexico, as well as pioneering recordings by African-American performers and composers, hillbilly and old-time music, cakewalk and rags, a group of unusual custom Blue Amberol cylinders recorded in Tahiti and German comic skits.

TWELVE

Berliner, Johnson, Seaman and Jones

> 'Those having a Gramophone may buy an assortment of Plates – comprising recitations, songs, chorus and instrumental solos or orchestral pieces of great variety. Collections of Plates become very valuable, and whole evenings may be spent going through a long list of interesting performances.'
>
> Gramophone operating instructions, 1898

With Volta and Edison drawing battle lines over the graphopone and phonograph, a third combatant entered the field, taking an entirely different approach. His name was Emile Berliner. He arrived in the United States on the ship *Hammonia*, on 11 May 1870. On the ship's manifest, the 19-year-old from Hanover described himself simply as a clerk.

He initially worked in Boston and Washington, DC, then took various low-end jobs in New York City while studying physics at the Cooper Union. His primary obsessions were electricity and sound, and by the tail-end of that decade he had learnt enough about acoustics to casually invent a microphone. Indeed, it was really rather an important microphone and, after he was awarded a patent in 1877, the rights to its manufacture were snapped up by the Bell Telephone Company, who promptly gave Berliner a job. Suddenly, the young, erudite German had money and time to focus on his passion for invention.

While living in Washington he had come across an example of Scott de Martinville's phonautograph at the Smithsonian Institution. He was immediately intrigued. He felt that the smoke-blackened surface, on which the phonautograph carved its wavy lines, also had potential to reproduce sound.

He started out by making his own version of the phonautograph. His was a rather thin and weedy cousin of the original, with only a narrow drum on which to trace its recording. It was literally only able to record a millisecond or so, but this gave him a bite-sized imprint of sound to experiment with. The next problem was how to reproduce it, which he did by having the soundbite photoengraved, or etched, onto a thin strip of metal. This he then wrapped around the same cylinder, rotated it, and let the recording travel back through the stylus, so transforming recorder to reproducer. He soon decided that this tiresome back and forth – removing the recording, having it photoengraved, then wrapping the metal strip back around the cylinder – was all far too much trouble, and so he began experimenting with the disc design that would ultimately cement his name in vinyl lore.

His first discs were made of glass covered with lampblack on a film of linseed oil. The recording stylus was underneath, carving upwards into the underside of the disc so material carved from the surface would fall away rather than clogging up the mechanism. A report into his work that appeared in *Electrical World* in late 1887 describes a device driven by a weight box that was able to record four minutes of sound on an 11-inch glass disc spinning at 30rpm.

Spurred on by the new Tainter and Bell graphophone that was doing the rounds, Berliner shared his findings with a patent attorney named Joseph Lyons in April 1887. By this time he was mixing oil with the lampblack, making a kind of fatty wax, which could then be copied in metal using photoengraving and played back on another device. He tried to demonstrate how *his* machine recorded sound in lateral movements, from side to side, rather than the hill-and-dale (or phono-cut) method used by his rivals. This, it was felt, would be the foundation for his new invention. He christened it the 'gramophone', a name that drew on Scott de Martinville's old phonautograms, and from the Greek *grámma*, meaning 'something written', and *phōnē*, meaning 'voice'. His machine could etch the human voice.

While Berliner waited for the patent to come through, he turned his attention to the method of reproduction – managing to create a negative matrix from the lampblacked disc, which could then be used to make zinc copies. Next he sought to eliminate all that photoengraving. After lots of trial and error, he succeeded by a method where a polished zinc plate was covered with a solution of beeswax and benzene – the evaporating benzene left behind a thin layer of wax. This was much better. The recording now cut into the wax, exposing the metal surface beneath. Dousing the disc with a solution of chromic acid then etched the grooves into the exposed metal. If you seek out an image of an early Berliner disc machine, you may see a strange glass jar with a long anointing beak. This pointy bird[22] was there to anoint the disc with alcohol, which helped prevent dust from sticking to the carving stylus. It was just around now, in May 1888, that Berliner gave his first public lecture and demonstration at the Franklin Institute in Philadelphia.

The experiments continued. The following year he tried using celluloid instead of zinc for his copies, but found them to wear too quickly, and by the summer of 1889 he was making discs from vulcanised rubber. It was these strange rubber discs that would be the first to go on sale. He was also keen to get rid of the hand-cranking method, and so began searching around for someone to help build a reliable spring mechanism, while one of his assistants began working on a coin-operated version.

Berliner interrupted research for a trip to Germany. During this trip he demonstrated his budding machine

---

[22] Self-indulgent reference to *The Man with Two Brains* (1983), a comedy starring Steve Martin in which he gifts his girlfriend a book of poems by fictional one-armed poet John Lillison, before reciting the ludicrous poem 'Pointy Birds'.

several times, attracting the attention of Waltershausen toy makers Kämmer & Reinhardt. The upshot of this German sojourn is the rather pleasing fact that the first ever commercially available modern-looking disc-shaped records were a single-sided 5-inch disc specifically designed for a toy gramophone and an even smaller 3-inch disc that worked with a talking doll. The sound quality of these hand-cranked machines was such that they were sold with small rectangular labels bearing the words that were being said. During their lifespan, according to writer and researcher Raymond R. Wile, author of *The Edison Discography* (2008), they were available in rubber, celluloid and more expensive zinc.

Fred Gaisberg met Emile Berliner during this experimental phase – some put this meeting as happening in 1891, although Gaisberg wrote about it being closer to 1893. Gaisberg was working with vaudeville veteran Billy Golden, whose 'Turkey in the Straw' had been a huge bestseller for Columbia in 1891, when Billy suggested they go check out what the competition was up to.

They visited Berliner's laboratory on New York Avenue, where the enthusiastic German was pacing up and down in front of the latest iteration of his machine. Soon he had Golden strapped up to a kind of muzzle with a long tube that led directly to the recording diaphragm, and Fred sat at a piano – its sounding-board boxed with another wider hose connected to the diaphragm. With a barked 'Go!', Berliner began hand-cranking the recorder.

Gaisberg was fascinated by this first glimpse of a completely new way of recording. Here the carving instrument moved laterally within a groove that stayed a uniform depth, rather than the hill-and-dale method he was used to. As he watched, Berliner took a bright zinc disc from the recorder and plunged it into a bath of acid. A few moments later, or so it seemed, he took the washed and cleaned disc and placed it on a reproducer and, again hand-cranking, played back the recording. Writing many years later, Fred recalled being spellbound by the 'beautiful round tone'. And he was impressed enough that he begged Berliner to give him work once the new machine was ready for business.

When Berliner felt able to launch, he transferred all his patents to the new United States Gramophone Company. His growing list of employees included William Sinkler Darby, Joseph and Zip Sanders, Raymond Gloetzner and Werner Seuss. Next he needed to amass a selection of material. Fred began working out of the offices at 1410 Pennsylvania Avenue, gathering a roster of performing artists, arranging recording sessions, playing the piano, as well as washing out acid tanks.

Berliner joined forces with a brilliant, single-minded engineer named Eldridge R. Johnson, who ran a machine shop in Camden, New Jersey. When hand-cranked, the optimum speed for a record was considered to be around 80 revolutions per minute. So Johnson was tasked with creating a reliable spring motor that could spin gramophones between 70 and 90 revolutions per minute. Two of the most common components available at that time were motors that ran at 3,600 revolutions per minute, with a gear with a ratio of 46:1. This resulted in a speed of 78.26 revolutions per minute. So the reason that 78 would become the dominant speed for several generations of record buyers is simply down to components available at the time. While the first gramophones on sale were hand-powered, the Baby Grand Gramophone, which came out in 1896, was the first to be powered by Johnson's spring motor, transforming the gramophone into a serious contender.

At the start, Berliner discs were all 7-inch. They lasted about two minutes and were still hard, vulcanised rubber with a name and date stamped in the centre. The quality of reproduction was really rather poor and, unlike many of the cylinder machines on the market, there was no way for buyers to record themselves. However, they had loads going for them. They were loud and relatively cheap. Hill-and-dale cylinders needed to keep the reproducer and stylus from jumping out of the grooves, whereas these discs had deep grooves that pulled the stylus and tonearm across the face of the disc. Cylinders needed boxes, labels and were very fragile. This disc of hard rubber was difficult to break, there was a nice space in the centre for the name of the artist, you didn't need any kind of box, and stored upright they hardly took up any space. The fact that many 'plates', as they were called, could be produced from a single master opened up more profitable vistas. And now artists wouldn't have to endlessly re-record performances.

Gaisberg had quickly brought together recording artists, offsetting the more expensive, established names with cheaper, local talent that he found on street corners and in bars. The most popular included Irish tenor George J. Gaskin, comedian Dan Quinn, and local monologists George Graham and his sidekick John O'Terrell. The wholesale cost of the gramophone was $3 per machine, $1.50 per dozen discs. Within a year they had sold around 1,000 machines, and closer to 25,000 discs, costing 60 cents each.

Even so, rubber just wasn't cutting it. Berliner turned to the Duranoid Company, which made electrical parts out of a shellac compound. He sent them a nickel-plated stamper, and the company returned a pressing that was, in comparison to rubber, more than a little bit awesome. Very soon, all Berliner discs were being pressed by Duranoid.

Berliner seemed to be in a good spot. He had an improving invention and, according to collector and expert Allen Koenigsberg, author of *The Patent History of the Phonograph* (1990), a strong patent that specified the self-driven, zigzag movement of the device, which became the rock on which Berliner was able to seek and secure new investors. He set up the new Berliner Gramophone Co with William C. Jones, and he sold licences to Frank Seaman, who formed the New York Gramophone Company, selling records and players in New York/New Jersey. In 1898, more than 700,000 Berliner discs were sold. Recordings were soon being made in both Washington, DC, and Philadelphia, stampers were being stamped in the capital, pressings were being pressed by Duranoid, and sales were handled from New York. However, Berliner was about to be hit by a wave of in-fighting, espionage and back-stabbing.

The aforementioned Frank Seaman is remembered as a particular troublemaker at this point. Although he had exclusive rights to sell and market Berliner's gramophones, he was rather hacked off with the terms of the deal he had struck. The Berliners, for their part, were hacked off with him for handing rights to one 'National Gramophone Company', when their initial agreement had been with him, and him alone. The rift widened as the rights to sell gramophones in Europe seemed to be about to come up for grabs.

While Seamen was annoyed about his deal with Berliner, the Berliner gramophones were also under constant attack from Columbia/Graphophone, who argued that Berliner was infringing the Bell-Tainter patents. In the midst of this already confusing situation, Berliner manufacturer Eldridge Johnson, who had been working rather secretively to improve the motors, sound quality and duplication of the Berliner discs, formed his own new company so he could continue making gramophones for the European market. Predictably, this attracted yet more angry attention from the graphophone lot.

All of a sudden, there was another newcomer, in the wonderfully named Zonophone (early on also styled as Zon-O-Phone). But Zonophone was in fact Seaman again, simply selling rebadged Berliner machines until he was told to stop. Then it resurfaced again as a coin-slot version of the Berliner gramophone, before in the middle of 1899 Seaman began selling his own, US-manufactured gramophone clones, once again as Zonophones. In 1899 a unique series of red discs by 'Vitaphone' appeared on the market. If you followed the paper trail, these Vitaphones turn out to be the graphophone group, all part of their assault on Berliner. Meanwhile, graphophone's first official foray into disc manufacture – the so-called 'Climax' discs, which were made by Globe Record Company – actually turn out to be the work of Eldridge Johnson.

Johnson by now had perfected a process of recording that would be familiar to record makers today – cutting into thick wax blanks, a metallic negative added to the surface by electrolysis, and the resulting negative used for stamping out copies. However, before he had patented this process, it was nabbed by a former employee from Berliner's laboratory named Joseph Jones. When it was granted a few years later, it was immediately purchased by the American Graphophone Company, who began manufacturing their very own 'Disc Graphophone'. So by the start of the twentieth century, the Berliner/Johnson lot were in the galling position of essentially abusing the patent covering the recording method they had come up with in the first place, and were being hounded by the very company who were in turn abusing their patents.

This period of open warfare prompted all sorts of innovation. Banned from using certain names, in fear of stepping on each other's patents, or simply wishing to break new commercial ground, the record makers, names and brands that flourished and died like fruit flies were a hotbed of R&D, constantly testing and improving, launching ingenious new devices and add-ons, trying out secret methods of recording and reproduction, and, by long, protracted litigation, setting the stage for the coming decades.

Eventually, while Columbia/Graphophone/Bell/Tainter failed to stop the Johnson production line, they did stop him using the name gramophone. So he dubbed them Victor Talking Machines and, in 1901, Johnson and Berliner came together as the Victor Talking Machine Company. With all the players having a hold over the others, at last they agreed to put aside their differences, allowing open use of all patents.

By 1902 there were three dominant figures: Edison (cylinders), Victor/Gramophone (discs) and Columbia/Graphophone (cylinders and discs). And while we're here, let's just have a check-in on terms. 'Phonograph', as we know, was Edison's word. As far as he was concerned, the only true phonographs were Edison-made, but it had also become the general word for all record-playing devices. 'Gramophone' started out as the trade name for disc players invented by Berliner. This would eventually morph into the Victor Talking Machine Company, and so the word would essentially disappear in America. However, the company would thrive in Britain, where the word 'gramophone' would come to mean any kind of record player for several generations. Columbia, meanwhile, carried on using the words 'graphophone' and 'grafonola' for their machines before the former slipped off into dictaphone territory. Victor made its first 'Victrola' in 1906 – a machine that had the horn built in to the cabinet. Soon 'victrola' came to be used as the catch-all term for any machines with internal horns. And let's not forget the term 'talking machines', which remained in very wide usage for some time to come.

The Gramophone Company Limited was founded in Europe in 1898, by William Barry Owen (acting as agent for Emile Berliner) and Edmund Trevor Lloyd Wynne Williams (who provided the cash). We'll hear more about the first British studio in a moment but, for now, it's just important to remember that this was the seed for a successful recording empire. The Gramophone Company would eventually merge with Columbia Graphophone Company in 1931, to form Electric and Musical Industries Limited (EMI).

Don't let this summary lead you to think of Edison, Victor/Gramophone and Columbia/Graphophone as being the only players in town. They were just the biggest, who, for the moment at least, could wield power over any smaller competitors that appeared.

THIRTEEN

Coin Slots and Record Shops

'I must again complain of the treatment of Grangetown by the parks committee in the matter of music in our open spaces ... If we cannot have a band, I trust we shall be saved the misfortune of giving hospitality to the gramophone ...'

Letter to the Cardiff Parks Committee, June 1910

On the Boulevard des Italiens in Paris stood Pathé's Salon du Phonographe. This plush, opulent destination was launched to cater for the demands of France's fashionable musos, where they could sample all the latest audio delights for a fee. Step inside and rows of deep velvet comfy chairs were arranged alongside polished mahogany cabinets. Each was fitted with a listening tube, coin slot and dial mechanism. Patrons were handed a catalogue and bought branded tokens costing 15 centimes apiece. Once they had selected a title, they dropped in the token, dialled the number, and a few seconds later would be listening to their selected choice. This wasn't a fully automated process or a proto-jukebox. Beneath their feet was a cellar, with stacks of cylinders and dozens of staff, who would receive the orders, fetch the appropriate cylinder, pop it in place and play the sounds.

This cutting-edge establishment traces its origins back to a bistro in Rue Fontaine near Place Pigalle, Paris, run by Emile and Charles Pathé. During a visit to the annual Vincennes fair, they came across a travelling showman surrounded by a gawping crowd, demonstrating the latest Edison phonograph. Visitors were paying to hear the device play speech and music from the wax cylinders. Hoping to cash in on the craze, they immediately approached the showman. When he refused to sell his model, they imported another from England.

Soon, people were flocking to the café. More and more weren't content with merely listening to whatever the Pathé's had; they wanted their own. The Pathés, unable or unwilling to obtain a licence for the official Edison machine, instead opted for a strikingly similar model they had made by a local company in Belleville. At the end of that same year, 1894, the Pathé brothers had a small factory in the Parisian suburb of Chatou producing cylinder blanks. The success of the Paris emporiums saw another Salon launched in Brussels, Belgium. Dutch newspapers ran advertisements announcing another new concern, which opened on a Saturday afternoon in

February at Reguliersbreestraat 43. Again, you could purchase a token, take a seat, peruse the catalogue, which by now had several thousand pieces of music to choose from, enter a number, drop a token ...

In Britain, there was nothing quite so grand as the Salon du Phonographe. Before the first specialist record shops, there were coin-slot phonographs, travelling showmen, then mail-order businesses, or departments within musical instrument shops or large department stores. Many of the first record dealers were instrument makers, while others sold phonographs as a natural adjunct to pianolas and musical boxes. Norwich piano dealer Saul Salkind branched out into talking machines, while Harrods sold gramophones out of its piano department. A very early stockist in London was the Oxford Street stationers Parkins & Gotto. By the later Victorian period the shop had giftware to suit 'every purpose, taste and pocket', and in 1891 began selling toy gramophones. Musical instrument store Imhof's was the first to sell 'proper' gramophones, later becoming the first official His Master's Voice dealer and going on to design its own product lines, including a bamboo stylus.

The average high-street gramophone dealer in Britain could often fix your bicycle as well. As Bruce Lindsay notes in *Shellac and Swing!*, the fictional shop of Grubb and Smallways from H. G. Wells' *The War In the Air* encapsulates the kinds of businesses that flourished in the early 1900s. These establishments worked fine for the average Joe with a little spending money, but they didn't appeal to the upper-class Edwardian music fan, who preferred to know that his or her dealer had some musical expertise, and so would be more likely to frequent London's musical instrument shops and department stores.

Spillers Records in Cardiff, Wales, founded in 1894 by Henry Spiller, is reputed to be the oldest specialist record shop in the world.[23] The long-established musical instrument firm Barnett Samuel, which would form the seed of the label Decca, entered the fray in 1901. Flaunty's Phono Stores was opened across the road from the Woolwich Arsenal on London's South Bank in 1903. William Flaunty was already selling records to friends when he was blinded by an accident at work. So he decided to make his sideline full time, and with his wife and daughter became the first official Gramophone Company agent in the area, continuing to sell phonographs and cylinders until he retired in 1940.

In pre-First World War Britain, the industry centred on London's City Road, where machines and records were produced and distributed, or sold through third parties. This was a seasonal trade, luxury items that tended to sell well in the autumn and the run-up to Christmas, then dwindle to almost nothing in the New Year. When it came to the machines, gramophones sold well in prosperous, urbane areas, while the cheaper German imports did better in more modest areas. As summarised by Science Museum curator Victor Kenneth Chew, the machines' Germanic origins would often be disguised, sold under patriotic names such as the Baron, Monarch or Marquis. German firms such as Fritz Puppel would sell low-cost cylinders and disc players to

---

[23] D'Amato Records in Valletta, Malta, also claims the 'world's oldest' tag.

the British through agents. British firms would import German-made players and then put them inside British-made cabinets, while others would import all the individual parts, assemble them over here and call them British-made. Some German cylinder models available in 1906 included the Magnet, the Excelsior, the Angelica, La Favourite, the Sylvia C, the Pandora and L'Enchantresse, while disc users could choose the Angelus or perhaps the Baby Tournaphone.

By now, reports and reviews of new pressings and machines were becoming the norm not only in specialist press, such as the *Talking Machine News*, *Phonotrader and Recorder* or *Sound Wave*, but also in the dailies and broadsheets. In January 1909, the *Weekly Mail* reviewed Columbia's raft of new 'pantomime songs', where Mr Frank Lombard 'sings very distinctly and with much humour'; where Wilkie Bard's latest comes with 'amusing patter'; 'If I Should Plant a Tiny Seed of Love' is appreciated as a 'robustly sentimental ballad'; a violin, flute and harp trio performing 'The Herd Girl's Dream' is wonderfully attractive; and it notes that 'dancing people' will appreciate Columbia orchestra's popular two-step 'Poppies'. The high street at Merthyr Tydfil had its own Pathéphone Salon, and in 1915 the proprietor gave an *Express* reporter a tour of his latest selection of Edison diamond-cuts, boasting grooves that were 'considerably finer than those of this variety that are now on the market, thus giving a longer record for the value', and also including some fine examples 'of the difficult art of obtaining good results with a choir of voices'.

A case relating to the theft of 200 records in Camden in 1906 gives some insight into mechanics of the trade.

When perfect, records were sold to British buyers at 3s 6d. Imperfect specimens were sold at 1s or sent back to the works to be refaced, recoated or, if beyond repair, destroyed. Mail-order businesses normally had some kind of returns system for when the notoriously fragile discs arrived damaged. An Edison-Bell product that arrived damaged at this time was very much the purchaser's problem, but Pathé would allow you to return it as long as you made a secondary order of the same value.

In the first decade of the twentieth century, dealers had to contend with price wars and bullying from the production companies. Over this period the average price of a cylinder reduced by half, and the price of a popular 10-inch disc by three-quarters. Companies also policed the prices, so a dealer who had overstocked on a popular title, hoping to sell at a certain price, might then be ordered to give massive loss-leading discounts. A tongue-in-cheek advertisement from 1907, under the headline 'A Dealer's Lament', essentially begged customers to help him shift 5,000 Edison records priced at 1s 6d before 8 August, when they would be reduced to 1s each: 'Waste no time or you may save your money.'

Back in the States, Edison dealers would generally peruse the *Edison Phonograph Monthly* for new releases and formats, as well as any new players coming to market. The 1905 April advance lists included the 'serio-comic' song by Harry MacDonough, a medley from the Edison Military Band, and a preacher's sermon on Adam and Eve, including singing and responses from a congregation. Wholesalers (jobbers) and dealers could then make orders for that month's stock. The company would send out sample records too, so jobbers could decide what they thought would sell. The Edison Gold Moulded Records sold in the US at 35 cents each, but dealers could also order the louder, larger and more expensive Concert format, which cost 75 cents.

Coin-slot operators had their own catalogue mailed to them. By 1905 the company made two types of coin-slot machine – the battery-powered Windsor, costing $80, and the direct current Majestic, costing $90 – that were sold alongside two spring-motor machines, the Bijou and the Excelsior. These grand-sounding names masked a frustrating experience for coin-slot operators. Even the company's own bulletin from March 1905 conceded that installing coin-slot phonographs, and keeping them going, required more time and effort than the average dealer could spare. Indeed, they announced a new trouble-shooting department, specifically created to keep the machines spinning.

There were hundreds of phonograph dealers by the 1890s and early 1900s. The hugely influential founder of American Decca, Jack Kapp, started his career at his father's Imperial Talking Machine Shop in Chicago, working at the shop after school, gaining a reputation for knowing the catalogue number of every record on the inventory. Other noted names included Henry Babson in Chicago, Lyon & Healy, the Eastern Talking Machine Company, Grinnell Brothers, Victor Rapke, C. J. Heppe & Son, and McGreal Brothers.

Into the 1910s and buyers might be treated to in-store 'Curtain' or 'Turntable' tests. These would have two rival machines playing from behind a curtain so listeners could decide which had the better sound. Turntables on turntables were even set up in the larger stores, so multiple rival machines could perform from exactly the same spot, giving a fair test of the sound reproduction. The machines would be maintained, operated and inspected by their respective dealerships to guard against any skulduggery. According to the City of London Phonograph and Gramophone Society, Edison's disc machines routinely won when up against machines and records from Victor, Brunswick or Columbia.

A famous name among American collectors was a San Franciscan called Peter Bacigalupi, who became the leading Edison wholesaler in the western states. The nameplate of Bacigalupi and Sons was affixed to thousands of phonographs, pianos and slot machines, and by 1906 he had three establishments in San Francisco – a wholesale house on Mission Street, a penny arcade and retail phonograph outlet on Market Street, and another penny arcade at the Bella Union Theatre, Kearny Street.

This all changed on the morning of 18 April when he awoke with a start on his bed like a 'bucking bronco'. When the earthquake ended he rushed downtown, finding buildings on their side, others looking like they had been cut in half with a cleaver. He hurried towards his store – slow going, as most people were struggling in the opposite direction, often lugging heavy valuables. 'I saw more talking machines in that one day than I believe I will ever see all together again in one time,' he wrote.

On reaching 6th Street, he was confronted with a wall of flame. He found his store, ran through the broken shop window, and began moving what he could towards the back, away from the fire, still thinking it might be saved. On reaching the second floor, the phonograph saleroom, he was astounded to see every record standing on its shelf in perfect order. 'This was the greatest wonder to me of all – to think that the pianos had been thrown down, and records, which stood by the thousands on our shelves, had not moved.' When he reached the roof he saw how hopeless the situation was. He descended the fire escape, and retreated to the penny arcade, where he sunk into an office chair surrounded by employees, all trying to cheer him up as he watched his wholesale house burn, followed by the penny arcade and the Kearny Street place. Two days later, with the fire still burning in some parts of the city, he secured new premises.

Of course not everyone could afford to buy their own record players, instead having to make do with first-wave turntablists and public dances. Early DJs in Britain didn't tend to nod heads and raise hands as climactic sections dropped. No, they tended to be so delighted the machine was working they wouldn't dare touch it. In November 1910, a grand concert at the Seaman's Institute at the Barry Docks, Wales, saw one Mr Alexander give a phonograph recital, just before some 'interesting lantern pictures'. In March, a reverend visited a Carmarthen workhouse, playing inmates a selection from eight discs that brought forward 'storms of laughter'. A fraud case in 1911 heard how a 21-year-old tailor's presser had invited friends up to his London flat to enjoy selections from his gramophone. And at a New Year's party in Port Erin, Isle of Man, 'great pleasure was derived from Mr Frank Page's massive Pathéphone'.

As machines grew louder, they appeared in stations, lobbies and public spaces. In 1910, large crowds enjoyed music supplied by a Pathéphone Majestic in Roath Park, Cardiff. Not everyone approved. An angry letter sent to the Cardiff Parks Committee read:

> I must again complain of the treatment of Grangetown by the parks committee in the matter of music in our open spaces. Last Tuesday evening a 'programme of vocal and instrumental concert' – such was the announcement – was promised for the edification and entertainment of the people of this district. Their credulity was, I am sorry to say, imposed upon, as they attended in large numbers to hear the capital-lettered MAJESTIC PATHEPHONE discourse its worthy music-hall and other airs. They are not likely to be again led astray, and will in future be wary of the intentions of those who cater for their enjoyment in this respect. I should have thought the parks committee might find some more profitable outlet for their energies than advertising, either under their own or other people's auspices, concerts of this description.

An electric gramophone was mounted on a platform overlooking the cavernous hall of London's Science Museum in the 1920s. These daily demonstrations were popular, with some people coming specially to listen to the symphonies and concertos. Presumably this sometimes caused overexcitement, as in 1929 museum attendants were told to immediately terminate any demonstrations if visitors became 'unruly'.

Tubes and Tone Tests

> 'Everywhere people were held spellbound by this daring test of tone re-creation. The ear could not distinguish the original from Edison's re-creation of it.'
>
> Advertisement, February 1916

During the period between the turn of the century and the start of the First World War, 70 different kinds of mainly needle-cut discs, plus another 30 types of cylinders, came to market in Britain alone. Cylinders were not inferior to disc records in terms of sound quality, but discs had fundamental advantages over their tubey forebears, which made them more adaptable for the long game. Even so, cylinders did not go quietly, and some of those produced in the first quarter of the twentieth century were not only relatively tough but could boast some of the best sounds available to the interwar generation.

The Edison company had a big problem to overcome. Berliner's discs from the outset were ready to reproduce. A method to upscale cylinder production became of vital importance. This occurred in 1901, with Edison's Gold Moulded, so called because of a gold vapour given off by electrodes used in the process. Here sub-masters were created from a gold master, and around 120 to 150 cylinders could be produced from a single mould. The new Gold Moulded records were black, pressed in a harder, tougher wax, with bevelled ends that bore the title and artist. Columbia too had its own moulded cylinders, and the marketeers fought it out, each claiming their own to be tougher and louder. An advertisement for Griggs' Music House, Des Moines, Iowa, made it abundantly clear which side of the road they were on: 'If any record were as good as the Edison Gold Moulded Record, or if any other Record were so good as to be almost as good, we would sell that other Record, but as yet we are unable to find that other Record. Until we do we will sell Edison's Gold Moulded Records exclusively.'

This became the battleground of the format wars. Alongside the cylinder versus disc campaign, there were constant skirmishes of cylinder on cylinder and disc on disc. Columbia, for some time, put out most popular and comic songs on both formats, leaving their more upmarket 'Grand Opera' series to appear on disc only. While discs were making gains, cylinders could still lay claim to better sound and were by now much cheaper. The larger Edison Concert cylinders from 1899 had cost

$4 apiece. Edisons in 1905 cost 35 cents, the Columbia equivalents just 25 cents.

The high tide of cylinder popularity in the United Kingdom was around 1906. The landscape was dominated by Edison, Edison Bell,[24] Columbia and Pathé. Several new names joined the flooded market, including Rex, Sterling and London Popular. Sterling cylinders arrived in a wave of publicity in 1906, doing well until they were price-matched by Edison, and then all but killed off by the Clarion cylinders, which arrived in 1907 costing even less.

The two-minute wax cylinder could not compete well against discs, which could offer up to four minutes of runtime. Edison's response was the Amberol, in November 1908, with its even harder wax and even finer grooves that doubled capacity to four minutes. In a classic piece of Edisonian hokum, the story was put about that the new tough black Amberols had a secret ingredient, known only to the great man himself who, like some grumpy Willy Wonka, would come out at some point in the production process and add the ingredient to the mix.

---

[24] The Edison Bell Phonograph Company was formed in 1892 and began life holding the distribution rights in Great Britain. There was a lot of tension with their parent (American Edison), who kept selling US-made machines in their territory. They also had skirmishes with a sewing-machine salesman from Manchester named James E. Hough, who had formed the company Edisonia. These eventually merged to become the Edison Bell Consolidated Phonograph Company and, when the Bell-Painter patents expired, they would move into both distribution and manufacture of machines and records, eventually becoming one of the major global producers of cylinders.

The old two-minute cylinders became known as Edison Standards, the four-minute records as black wax Amberols. In a bid for more high-class content, the company had been courting renowned opera singers for a couple of years, recording them in a 17th-floor studio on Fifth Avenue. Having been constrained by the two-minute run-time of the Standards, there were high hopes for a new series of Grand Opera Amberols, priced at one or two dollars each. They put out more than 150 of these, but they never sold well. The last hurrah of the Edison cylinder, and the most avidly collected, were the Blue Amberols launched in 1912, boasting a four minute and 45 second run-time, with a surface layer of 'indestructible' plastic celluloid that could be played '3,000 times without wear', and tinted a distinctive blue.

The Edison Company had been stamping its feet, attempting to enforce moulding patents across the globe. While justified, the pouting was badly timed. The patent headache meant that many manufacturers, suddenly barred or at least discouraged from making cylinders, turned all their energy towards discs. All sorts of formats, speeds, sizes, cuts and colours were tested. Gramophone launched its single-sided 10-inch shellac disc in 1901, and 12-inch discs two years later. Pathé, which abandoned cylinders for good in 1906, began making phono-cut discs that started in the middle and used a sapphire instead of steel needle. Supposedly unbreakable but nevertheless doomed, Nicoles came in a fetching brown-orange celluloid on card, the short-lived Neophones were white, the 1909 advertisement for new Millophones called them 'THE RECORD of the season' (but they still vanished), while Musogram launched a series of longer-playing but equally short-lived Marathon discs.

The German label Odeon produced a roster of double-sided titles in 1904 – the 7½-inch Standards and the 10¾-inch Concerts. An Odeon advertisement hollered 'a different selection on each side', demonstrating this step forward with an image of a disc being held before a mirror, revealing its grooved backside. The words 'One Odeon Disc Record' were double underlined, emphasising that each held two, yes, TWO tracks from the same artist. The point was further hammered home with a list of 14 reasons why these discs were superior to competitors: including the fact you got two, yes, TWO records in one, postage costs were halved, and let's not forget that the grooved underside held better to the turntable felt.

More German discs flooded in behind such monikers as Jumbo, Favorite and Homophone, forcing prices down. In the UK, a standard price for a double-sided 10-inch record had been 5s. Within five years this had come down by half. Even the Gramophone Company, which appeared to stick its nose up, preferring to adopt a reassuringly expensive attitude, quietly sold cheaper discs under different brands such as the Twin label. The race to the bottom culminated in James E. Hough introducing the 1s 6d Winner; Gramophone launching its Zonophone Cinch (1s 1d) and Columbia the Phoenix. By the time war broke out, some records could be bought for less than a shilling.

Pathé continued to do things its own way. Cylinder customers could either go for the Standard or the larger, louder Salon. Pathé's new hill-and-dale discs could either be played on home-grown Pathéphones or, for gramophone users, via an inexpensive adaptor. Pathé had an unusual way of recording its masters, too

– performances were cut into large cylinders named Paradis, before being transcribed to wax 78 discs, from which they made a matrix.

Pathé weren't the only hill-and-dale discs in town. Britain was treated to Neophones in a wave of publicity in 1904. These were very light, laminated onto a cardboard base and, by all accounts, terrible. As with the Pathé discs, you could either buy a specialist player or a Neophone adaptor. They also produced some more expensive 10s 20-inch discs that could play for eight to ten minutes. Although this was the longest available to the market at the time, and so can be seen as a short-lived precursor to the long-playing record, it too sounded terrible.

You can follow the struggle of cylinders and discs in America through the pages of trade rag *Edison Phonograph Monthly*, which was sent out to wholesalers and dealers across the States. The pages are filled with bland-washing headlines like 'Another Great Month', with success stories from the entertainment and coin-slot arms of the business, alongside catalogues of new releases, new phonograph designs and lots of bad-mouthing of anything disc-shaped. Reports from all parts of the country tell the same story, of the 'great increase in popularity of Edison goods', and dealers abandoning discs in favour of Edison phonographs and Gold Moulded cylinders.

Despite this public-facing attitude towards the flat upstarts, behind closed doors Edison's lab was experimenting with them. Chemist Jonas W. Aylsworth had been working with the company since 1890. The story goes that Aylsworth and Edison's veteran recording expert Walter H. Miller saw the need for a disc and began working in secret. Aylsworth and Miller were dismayed when Edison found out what they were doing, and even more dismayed when he decided to take charge. Vigorous denials of anything circular continued, until a pending machine was finally announced in July 1911, at the fifth annual convention of the National Association of Talking Machine Jobbers in Milwaukee, Wisconsin.

To head off trouble from Victor's patent lawyers, they stressed that the new machines would be based on Edison's original 1878 British patent, and they decided to pitch for the quality end of the market. In Britain, members of the Northants Talking Machine Society were the first to enjoy a demonstration of these new Edison Diamond Discs, as they were known. They were summarised as having a 'good tone', though not very loud. (In Britain, Edison's discs would suffer from the ban on all non-essential imports, and after the war were simply too pricey. In 1919 an Edison disc cost around 8s 6d, more than three times the price of popular competitors.)

The Diamond Discs sounded excellent. They had been recorded in a deadened studio, cutting out ambient noise, and they had less surface noise than some of their competitors. To spread the word, they were put through a long-running series of live, highly choreographed, well-publicised and very popular 'Tone Tests', in which star singers would perform alongside their own discs, and the audience would be challenged to tell the difference. These weren't the first public demonstrations – in 1907, HMV had hired the Albert Hall in London to spin some discs aboard the new compressed air 'Auxetophone' model, for example – but the way they were pitched caught the public imagination.

The Tone Tests kicked off in October 1915 at the Panama-Pacific Exposition in San Francisco. The machine, specifically in this case a Chippendale C-250, would play a piece of music all the way through, with the singers either singing alongside or just moving their lips. Sometimes they'd plunge the stage into darkness during the live test, or mask them behind curtains. The recordings used for the demonstrations were carefully selected, usually ones with fairly sparse backing. Singers too were chosen because of their ability to imitate the sound of their own recordings. Operatic soprano Anna Case performed a high-profile Tone Test at Carnegie Hall, many years later explaining that if she had sung at full volume, it simply wouldn't have worked, so she reined herself in and tried to sound like her record. Interviews with some of the original Tone Test artists also appeared in the 1930s *Hobbies* magazine. They too admitted to the mild subterfuge of singing quietly or adopting a more nasal voice. In any event, on countless occasions, the discs performed well, fooling the audience.

Between 1915 and 1920, the company sponsored more than 4,000 Tone Tests. Some took place at leading venues in front of large audiences, but the majority were modest, bringing this strange combination of live and lip sync to local stores and town halls. A photograph in *Edison Phonograph Monthly* in 1915 shows a test being set up in a piano and phonograph store in Toronto.

Charismatic, robust baritone Arthur Collins, sometimes called the 'King of the Ragtime Singers', was approaching the end of his career when he took part in one of the Tone Tests. Although he survived the ordeal, a telegram sent to Edison incorporated in October 1921 read: 'Collins painfully injured tonight in Medina by falling through trap door during dark scene. Doctor took fifteen stitches in scalp. Have brought him to Cleveland. Do not notify family. Will wire again when doctor here sees him in the morning.'

In 1906 the Victor Victrola Talking Machine was launched. Here the horn was placed inside the cabinet, which in terms of reproduction was inferior to external horns but it was enormously popular because it was stylish and unobtrusive. Even though the cylinder market was in freefall by 1909, Edison decided they needed an answer to the Victrola, and came up with his own internal horn cylinder phonograph called the Amberola, designed to play the four-minute Blue Amberols or Columbia's Indestructibles. The Amberola 50 of 1911, for example, boasted a wood grate and a diamond stylus. You'd raise the lid, pop your Blue Amberol onto the mandrel, wind the crank until you felt the belt start to give some tension, flip the switch to start it spinning, move the diamond stylus over, and lower the lever. These, remember, were purely mechanical, so any volume control came from the gauge of needles you were using – soft, medium and loud tone – which, of course, you had to change after every play.

While Columbia had responded to Edison's black Amberols with its Indestructible Cylinders, these were to be its final innovation in the area. It gave up making its own cylinders in 1909, abandoning them altogether in 1912. When Pathé started making their hill-and-dale discs in 1906, they promised to keep cylinders going alongside, but soon left them behind. The United States Phonograph Company stopped making its Everlasting cylinders in 1913. Edison soldiered on in America but was soon forced to close the European production plants. From that point on, all cylinders originated from the factory in West Orange.

The Blue Amberols continued their sombre final march, in packaging plastered with Edison's name, face and patent information. They genuinely could be said to have the best sound available at the time. The constant surface speed of needle on groove was superior to the inner-groove distortion that occurred with discs as the surface speed reduced. Hill-and-dale diehards could also argue that the vertical cut groove was fundamentally superior to the lateral cut.[25] It certainly was a simpler prospect in the days of acoustic recording. As described by Gelatt, a sound engineer working with lateral-cut discs had to guard against sudden loud noises that might cut right through the groove walls. Vertical cuts didn't have this problem. The combination of surface and precision-ground stylus also meant less noise. Even though the Blue Amberols continued to be made, towards the end they were being dubbed from Edison's Diamond Discs. That's a bit like ordering an original Betamax video for your cherished video player, only to discover that it's been pirated from a VHS copy. This is the reason why collectors are particularly passionate about the earlier examples. Although Blue Amberols were as hardy as grooved media could be back in the day, over time the celluloid is susceptible to shrinkage.

---

[25] Not really true. Many would argue the dynamic range of hill and dale was limited by the depth of the record, whereas lateral cuts had lots of elbow room.

The coming of commercial radio would deal a death blow to Edison's cylinders. Record sales in all formats plummeted. The price of Edison cylinders was slashed, dealers left the trade, and for years the only way to get hold of any copies was direct from the factory. In the summer of 1929, the last Blue Amberols were released. The Edison company's phonographic division ceased all manufacturing on 28 October that year.

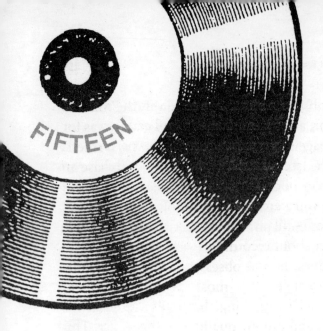

Studios, Scouts and Engineers

> 'We didn't know it at the time, but my boys and I were frontiersmen, pioneering a new science destined to grow into a tremendous industry.'
>
> Paul Whiteman, 1948

The First music mogul was a dashing Italian cavalry lieutenant. On a visit to Paris, Gianni Bettini had fallen in love with an American socialite named Daisy Abbott. He quit the military and followed her across the Atlantic. Once they were married and settled, he set his energy towards invention. And for Bettini, the moment he heard about the phonograph, it existed solely as a means for recording and preserving music.

Bettini was obsessed with music. He was the son of a renowned tenor, and his first patent application was for a mechanical page turner, designed to help performers. Sometime in the late 1880s, he managed to purchase an improved Edison phonograph. This was right back at the start, when you weren't officially able to buy them at all, as they were still available only under lease. He immediately set about recording.

From the outset, he was obsessed with achieving the best sound possible, the most perfect record of the person's voice, something that truly captured the essence, timbre and unique qualities of the singer. This all-business phonograph just wasn't up to the job. And he concluded, after some experimenting, that it came down to the recording diaphragm – the fact that it had just a single point of contact with the cutting stylus. He began experimenting, making a more flexible diaphragm, attaching the stylus with multiple 'radial spurs'. The perfected version of these delicate spidery prototypes would eventually be sold as attachments, 'Bettini's micro-reproducer', designed to be fitted to Edison machines to give users 'a true mirror of sound'.

The Bettini household became a popular destination for members of New York's fashionable music scene, and Bettini would cajole and charm many of the most celebrated opera stars of the day to record for him. Not only did he slowly amass an archive of talent, but he even invented a system of duplication. The master cylinders were kept at his studio on Fifth Avenue, and he perfected a machine that could 'transcribe' copies for sale – like a kind of very basic double tape deck. However, keep in mind that the 'masters' in this case were still only the original cylinders. And just like other cylinders of the day, they wore out. Each time he took a copy, the master would be less able to produce more copies.

Bettini didn't pay the friends who sang for him, which is perhaps central to why so many did – it was more a labour of love, an informal agreement between friends. His collection attracted lots of contemporary publicity, later becoming something of a Holy Grail among collectors and aficionados. A report in *The Phonoscope* in 1896 described the collection as 'unequalled', with mouth-watering descriptions of voices captured with unrivalled richness and clarity, high notes that soared, bass notes clear of distortion. Bettini would eventually take his collection of masters back to France, where the majority were destroyed during the war. Over the years, examples have resurfaced, causing ripples of excitement among collectors and historians. One man named A. R. Philips found two caches – while on Navy leave in Mexico City in 1945 he found some in an antique shop, then 17 years later discovered some more in a Parisian flea market.

Bettini was perhaps one of the first true amateur audiophiles, who learnt, through trial and error, the variables that controlled sound quality – the thickness and shape of the diaphragm, the design and size of the horn, the importance of the cutting stylus, and so on. Despite the fact he invented his own means of duplicating recordings, his own stylus, and recorded an entire generation of performers, Bettini did not revolutionise the industry.

Sound recording can be roughly divided into four periods: acoustic (1877–1925), electric (1925–45), magnetic (1945–75) and digital (1975 to date). In the acoustic era, sound travelled down the horn, towards a diaphragm with stylus, which etched the sound waves onto wax cylinders or discs. Once you were ready to play back an acoustic recording, a reproducing point – usually a steel or sapphire needle – was affixed to an encased diaphragm called a sound box or reproducer, and this was attached to a tapering tube known as the tonearm.

The Victorian and Edwardian sound studio had no fancy tone controls to play with. Any adjustments were made by altering the performer's position relative to the horn, or by trying horns of differing shapes and sizes or diaphragms of varying thickness. During the acoustic era, sound engineers were only able to capture limited audio frequencies – from around 100 to 2,500Hz – which, as we've touched on, favoured tenors and baritones, banjos and brass. Studios gradually figured out that a wider range of frequencies could be achieved with thinner, more sensitive diaphragms, a sharper stylus and softer wax for the high notes. Most would have a large array of recording horns at their disposal – different sizes, shapes, thicknesses, tailored towards the performer. And at the same time, artists also learnt the ropes, becoming more practised at getting the best out of each set-up – altering their voice to suit the recording medium, changing their position depending on the passage of music, and the strength and volume of their singing. A reviewer on *Talking Machine News* described how some instruments were still proving 'baffling to the recorder', that pianos had once sounded more like banjos, and that the only option for violins were the Stroh instruments – these came with conical horns connected to the bridge, which created a harsh, rasping sound in the flesh but translated relatively well to wax. Scarcely a decade later and engineers were getting piano 'reproduction as close to perfection' as was possible. The reviewer describes a contemporary piece by French pianist Alfred Cortot, in which the 'exact tones of a grand piano were given without the slightest difference from the instrument ... save for the inevitable surface scratch'.

Oliver Read and Walter L. Welch, the authors of *From Tin Foil to Stereo: Evolution of the Phonograph* (1971), suggest that the New York phase of Bettini's career was brought to an end by the beginnings of commercialisation – and this we can witness through the lens of Fred Gaisberg, and in particular his exploits in Britain.

Gaisberg had helped set up the first of the disc-gramophone recording studios in America, above a shoe shop on 12th Street, Philadelphia, in 1897. At that time, the Berliner discs were still in their first iteration – the single-sided 7-inch records, playing up to two minutes of sound. There were no royalties or music publishers – singers were generally paid a $2 or $3 flat fee per song. Soon the company had studios in New York and Philadelphia and, on seeing how well Pathé seemed to be doing, it was time to set up in Europe.

In the summer of 1897, Berliner sent over a bold, quick-thinking legal assistant named William Barry Owen. Owen gathered a syndicate of investors, headed by Trevor Williams, and by the end of the year they were closing in on an agreement – the European rights to Berliner's gramophone for $5,000, using American-made machines and discs to begin with, but with the promise they could start producing home-grown recordings. The first meeting of the Gramophone Company took place at the Hotel Cecil in April 1898 – Williams was the chairman, Owen the MD. They ordered a starting stockpile of 3,000 machines and 150,000 American records while they looked for an expert to help set up a new recording studio.

Gaisberg crossed the Atlantic on the Cunard liner *Umbria*. He had recording equipment, a bicycle and a notebook full of instructions and letters of introduction. He arrived in Liverpool and made his way to London, the company's first address being just off the Strand at 31 Maiden Lane. While looking for lodgings, he immediately set about stocking up the studio with the tools of the trade – an etching tank, linoleum, cotton cloth, scissors, soldering iron, coal oil, acid.

This then was the first gramophone studio in London – a grimy smoking room of the old Coburn Hotel. To begin with there was one recording machine on a high stand, an array of long, thin horns, and an upright piano on a movable platform. While it might not have looked much, it was very close to the West End, and Gaisberg soon set about doing what he was good at – persuading local talent to record. One of the first was an American named Bert Shepard, who became an early star in Britain, remaking many of the popular comic recordings that had already been hits in the States, such as George W. Johnson's 'The Laughing Song'. Fred persuaded him to record and taught him many of the popular tunes of the day – simply playing them over and over, and writing down the lyrics.

The London studio had an open-house policy. Amid more well-known performers was an 'endless procession' of random players, whistlers and singers. One named George Mozart misunderstood how it all worked and stood before the recording horn in full make-up and costume. Stout was apparently the drink of choice at the studio and Fred was always astounded at the number of empties by the end of a session – apparently the average consumption of music hall singer Harry Fay was six bottles. A local publican would often bring along performers who had been drinking in his place, including comedian Dan Leno, and they always tried to keep the recording machines ready to go at a moment's notice. If Fred wasn't playing, many of the early records were accompanied by Amy Williams, the recording department secretary who doubled as the in-house pianist.

Soon the record-pressing machinery from America had arrived in Europe. Berliner's nephew, Joseph, was in charge of setting up a factory in Hanover but hadn't yet completed the task. As a result, the first Berliner discs in Europe were pressed in a large tent. Finished discs would arrive back in London about a month after the masters had been sent over to Hanover. And by the end of the year, there were several hundred shopkeepers and agents selling gramophones and records.

Fred's work continued overseas. With portable recording equipment that fitted into six packing cases, he and his moustachioed, pipe-smoking wingman William Sinkler Darby set off across Europe, recording performers in improvised studios, usually hotel rooms. They recorded in Leipzig in 1899, then Budapest, Vienna, Naples and Milan, then Paris, Madrid and Valencia. In Paris, Alfred Clark had gathered together a roster of locals to perform ahead of the arrival of Gaisberg and Darby, from café performers to opera stars, including a wildly popular baritone named Léon Melchissédec. After a brief return to Maiden Lane, they were sent off again, this time to Glasgow, Belfast, Dublin and Cardiff. Gaisberg would later cut India's first gramophone recordings from a makeshift studio rigged up in two rooms of a Calcutta hotel, and in Japan, he recorded more than 270 titles in a single month. As the military situation in South Africa had deteriorated, they even cut some of the first war records. This included an innovative atmospheric, descriptive record called 'The Departure of the Troopship', in fact written by the original 'Casey' Russell Hunting, and also 'the Absent-Minded Beggar', a fund-raising morale-boosting poem by Rudyard Kipling that had been put to music by Arthur Sullivan.

By now Fred's brother, Will, had joined the company, and Eldridge Johnson had perfected the 10-inch record, allowing for longer pieces, opening up more possibilities for recording opera. They departed on a two-week expedition to Russia, setting up a studio in St Petersburg, where they bagged tenor Leonid Sobinov, and many other singers from the Imperial Opera. By the end of the trip, another 254 new masters were on their way to the pressing plant in Hanover.

The Gaisberg brothers were sent to Milan during the Scala Season in early 1902. They hoped to record a new batch of opera talent ... and the pope. Operation record-the-pope didn't come together, but they did manage to record the last castrato singer in the Sistine Chapel Choir, one Alessandro Moreschi, in an ornate room of the Vatican Palace. During one of the final sessions a short-circuiting battery ignited cotton wool used in the packing cases. The fire caused a stampede of choristers, before they managed to control the blaze with overcoats, and the precious recording was saved.

The Gramophone scouts had secured that season's hottest tickets – for *Germainia*, in which a new singing sensation was holding audiences spellbound. The young superstar was Enrico Caruso. Suitably impressed, and having navigated Caruso's entourage, they approached him with the simple question: how much to record 10 songs? A go-between returned the next day with the answer: Caruso would sing 10 songs for £100, and he wanted it to be done in a single afternoon.

This was a staggering, eye-watering, unheard-of amount of money. Compare this with the earliest surviving royalty contract from the Gramophone Company archive, with the music hall star Albert Chevalier, dated October 1898. In that, Albert was promised 1s per dozen records sold – that's one penny per record. Fred, however, sent word of the terms to London, with his recommendation that they should go for it anyway, receiving the telegrammed reply: 'FEE EXORBITANT FORBID YOU TO RECORD'. And yet this was the ultimate scalp, a record signing, a first opportunity to capture one of the greatest musical stars of the day. Fred ignored the telegram, apparently deciding that he would simply guarantee the money out of his own pocket. He figured that to make back the £100, they would have to sell at least 2,000 copies.

There was no time to record until after *Germainia*'s run. They had a chance to test their hardware with another rising star, the opera's soprano Amelia Pinto, who signed a more modest but still very handsome contract worth £40. Then, on 11 April, Caruso sauntered into the Grande Hotel. The sessions would prove to be a huge success. He was no difficult diva, but an absurdly talented man, full of likeable confidence and good humour. Quickly, without fuss, he recorded 10 songs, including 'Studenti! Unite!' from *Germainia*. They finished, and Fred handed over his £100.

You can imagine how nervous the team must have been around those delicate, priceless recordings. The Caruso masters were so precious they transported them home by hand. The records were pressed in Hanover and arrived on sale in London, just in time for Caruso's May debut in Covent Garden.

They were an enormous financial success, quickly exceeding the 2,000 sales Fred needed to break even. In the US the recordings spearheaded Victor's new premium Red Seal line, and Caruso cut seven new titles in Milan in November, including repeats of slightly botched numbers from the first session. He made his debut at the Met in 1903, which effectively became his home turf for the next two decades. He would record exclusively for Victor for the rest of his life.

In the immediate aftermath of Caruso's first session, Gramophone began signing contracts left and right, with many stars in town for the Covent Garden season. Fred and William together either signed or recorded the likes of Adelina Patti, Francesco Tamagno, Feodor Chaliapin and Nellie Melba. Some took to the horn with ease, others found it tricky. French opera star Emma Calvé insisted on dancing as she performed, or would break off midperformance to ask if her voice was alright.

Many singers were far from impressed by the spit-and-sawdust surroundings of Maiden Lane. Today it's a place of pilgrimage – where the recording industry in the United Kingdom began. While the studio is often described as being in the basement, many experts dispute this, and argue that it was in fact located on the first floor. What is not in doubt, however, is that it was by all accounts a grimy, unimpressive venue for high art.

By the summer of 1902, Gramophone went more upmarket, launching their new 'Red Label' series and moving to new promises at 21 City Road. The tailor-made studio was located on the top floor, far away from street noise. A 1907 description printed in *Talking*

Machine News describes a variety of 'every sound-producing instrument' on hand, including organs, chimes, pianos, drums, horns and strings, alongside countless receiving horns of every size and shape. There were curtains and drapes to help soundproof sections of the room, while recording horns were placed through curtained windows in movable partitions.

Another reviewer from *Talking Machine News* would provide us with an account of the contemporary British recording studio. 'To get behind the scenes of the recording room,' he wrote, 'is a very difficult matter, for the secrets there are most jealously guarded.' His experience begins with an interminable number of stairs, leading to a bare apartment with a single music stand. There a tall engineer points at a long trumpet-like tube projecting from the wall. He's told to stand with his mouth about 6 inches from the opening, with the instruction that when the light shows, fire away. The engineer leaves the room.

'It was desperately uncanny.'

The light flashes, and he begins, in this case reading some poetry. Halfway through he makes a slip and the light signals again. From the other side of the wall comes the voice of the young man: 'Pity. You were doing very well. I've got another blank.'

So much for a human reading poetry. Duets, trios, quartets and choruses, bands and orchestras required more finesse. 'On stepping into the private domain of the recording expert we were at once sensible of a considerable rise in the temperature ...' This had to be maintained so that the wax of the blanks was at the right consistency for the recording needle. The blanks themselves were often kept in warmed cupboards around the room – and there was a machine for shaving spoilt blanks for reuse.

The recording machines had to be kept completely stable, any slight vibration not direct from the diaphragm would ruin a recording. At this time, most recording machines were powered by electric motors or weight-driven mechanisms. The diaphragm was almost always of glass – the 'substance most amenable to the action of the sound waves' – and a lever attachment fit it to the cutting point. Different sound boxes were used for cutting depending on the situation. An engineer might use one for the soprano, another for the contralto; a carving stylus that worked perfectly for a violin solo might be of no use for recording a string quartet.

'The wax blank having been selected, dusted to remove every particle of foreign matter and placed upon the turntable, the recorder gives the signal and the pianist in the next room begins ... The artist follows with the air and all the time the blank is spinning with the turntable. We watch it with a sort of fascination. Thin threads of composition curl up from the jewel point and are blown off as they rise, till presently the song is finished, and the vacant space on the wax has become covered with grooves.' After a successful take, the engineer would take up the wax, go over it carefully with a camel-hair brush, before packing it in a box with cotton wool to be sent to the factory.

Sound engineers would guard their cutting heads carefully. American bandleader Paul Whiteman worked with Victor's trio of engineer brothers Charlie, Raymond and Harry Sooey. The elder brother had passed down the secrets of the cutting head to his siblings, and now each had worked on their own improvements. The brothers kept their cutting heads in little leather boxes, which never left their sides, day or night.

Whiteman was at the epicentre of the dance craze of the 1920s. At the start of the decade, he began recording for Victor, and through years of trial and error, had figured how to get the best from acoustic recording. They found, for example, that ordinary drums just could not be made to record properly – timpani and snare were alright, but the moment the bass drum entered the mix, it created a fuzzed-up sound that impacted everything else. To begin with they tended to use a banjo instead as a kind of 'tune drum' to mark out time, and they reinforced the brass section with various combinations of trumpets and trombones. They also pioneered a full quartet salvo of saxophones when most studios only used single instruments. They would usually record two sides at a time, and through rehearsal and the process of carrying out test cuts, each required about two hours of preparation.

Whiteman cut the first version of George Gershwin's 'Rhapsody in Blue' in New York on 10 June 1924. While an average track might employ around eight or nine musicians, for 'Rhapsody in Blue' they used 25, including Gershwin himself on piano. 'We filed into a tiny, dusky room, barren, crowded and uncomfortable,' wrote Whiteman. In the middle stood an 8ft-high tower, made up of four ladder-like supports tapering to a narrow point, with four recording horns, four or five musicians gathered around each. 'My boys had to be athletes,' he wrote. When a solo passage was played, the musician would move up close to the horn and play directly into it. Then back out in a hurry, dodging out of the way of the next man rushing towards the recording pylon.

Nipper was a dog from Bristol. Born in 1884, he was probably a mix between a Smooth Fox Terrier or Jack Russell Terrier. His owner was a scenery designer named Mark Henry Barraud. After Mark died, the pet ended up with Mark's widow in Kingston upon Thames. In 1898, three years after the dog too had shuffled off, Mark's brother Francis painted a picture of Nipper with a slightly cocked head, listening intently to a cylinder. The popular myth was that the painting was inspired by an actual event, with Nipper listening to the recorded voice of his departed master. While this part of the story is untrue, it's perfectly possible that the image was inspired by the creature's genuine interest in and confusion caused by noises emanating from a phonograph.

In late May 1899, Barraud took what would become known as 'His Master's Voice' to the offices of the Gramophone Company in London. In the original painting, the horn was black. The artist wished to borrow a brass horn, so he could repaint and brighten up. It seems likely that he was also there in the hope that what was about to happen would happen. William Barry Owen loved the painting so much that he offered to buy it, on condition the artist replace the Edison cylinder machine used in the original with a disc gramophone.

Gramophone discs from this period were pressed with what is known to collectors as the 'Recording Angel' logo. Owen had no thought of turning 'His Master's Voice' into a logo or emblem; it was just an arresting image that he liked. Nevertheless, it was used on a company catalogue in December 1899, and while it's not entirely clear if it was Berliner or Johnson who first brought the image across to use in America, it was certainly Johnson who was instrumental in adopting it as a trademark for Victor. Barraud was suddenly busy with commissions for new copies for various corporate uses. Thousands of prints were produced in the UK, sold to dealers for 2s 6d, and soon Nipper and the slogan were fronting American-made discs and players.

In the earliest known version of the painting, Nipper is sitting on a highly polished piano-black surface, with clearly defined edges. Over the years, the painting and logo would be adapted, cropped and stylised, losing the reflected image of the dog and record player, and any hint of a defined surface; one urban myth has it that in the original, this polished surface was intended to depict a coffin – and in fact the coffin of the dog's master. As HMV became one of the most widely recognised and popular brands, it only fuelled a fascination with the macabre story, which was then endlessly repeated, refuted and unpicked. Whether it was originally intended to be a coffin is a moot point. It's certainly true that a dog sat on a coffin listening to its dead master's voice is perfectly in line with lots of other gloomy Victoriana of the day. And it's also true that the phonograph was linked with death from the very beginning. The writers of that first *Scientific American* editorial predict the strong emotion readers will feel at the thought of this new power to preserve the voices of loved ones. The idea of the preservation of a voice after death was a common trope in the phonograph's advertising copy. There were reports of the machines being used at funerals, including one sensationalist story of a preacher giving the sermon at his own funeral. James Baird McClure wrote in 1879: 'How startling also it will be to reproduce and hear at pleasure the voice of the dead! All of these things are to be common, everyday experiences within a few years.' A set of operating instructions from an early model gramophone in Britain ran: 'The Gramophone is intended and expected to be for the voice what photography is for the features, i.e. to secure an accurate and lasting record of voices.

British discs and players began using the Nipper and HMV branding after their US counterparts. When Captain Robert Falcon Scott embarked upon the *Terra Nova* expedition to Antarctica in 1910, he took with him two HMV 'Monarch Gramophones', loaned by the Gramophone Company, and several hundred 78rpm shellac discs, chosen to boost the team's morale. The expedition photographer even recreated the image, with one of the Siberian sledge dogs supposedly listening to the machine, which was widely reprinted under the headline 'His Master's Voice at the World's End'. The original painting would end up at EMI's Gloucester Place headquarters in London, while Nipper's remains lie beneath the rear car park of a bank in Clarence Street.

There was something of a seismic shift that took place when Caruso made those records in Milan. The publicity and the quality of the records helped transform the humble phonograph into something the serious music lover desired. Gramophone launched their premium-priced imprint for the very best performing artists. Vaudevillian whistlers and music hall maestros were not welcome here, thank you very much, only the noted tenors, sopranos and baritones of the day.

Depending on where in the world you lived, they would become known as Gramophone's 'Red Labels', or

Victor's 'Red Seals'. The first 'Gramophone Record Red Seal' disc appeared in late 1901. The first Red Seals recorded by Victor in the United States were of the Australian contralto Ada Crossley, in April 1903. Eldridge Johnson wasn't too worried about the eye-watering increased artists' fees, nor was he overly concerned about whether they made a profit. For one, he was confident they would in the end prove profitable, and he understood that their great value lay in aligning the brand with high class and high art. Newspapers and magazines ran enormous advertisements with Nipper the dog, alongside their latest Red Seal titles. The office building in New York had an enormous 50ft square light-flooded painting of Nipper on the roof. Five of Caruso's Milan recording were issued by Victor on the Red Seal label in March 1903. And Caruso's second session, in February 1905, produced five more. Together their success encouraged more stars to join the line, including Marie Michailowa, Marcella Sembrich, Marcel Journet and Emma Eames.

In 1906, Caruso recorded again. The Stroh violins, cellist and wind instrument players standing on chairs helped give the recordings a tin-can version of the full orchestra pit. These were louder, clearer than anything recorded before. And in March they recorded the first duet – Enrico Caruso and Antonio Scotti – which would become an international bestseller.

In the coming era of electrical recording, complex orchestration could be captured as never before. The Red Seal series became the pre-eminent classical music record label, led in particular by the recordings of three conductors: Serge Koussevitzky, Leopold Stokowski and Arturo Toscanini. Toscanini began his recording career in 1920 and nearly all of his recordings were issued on the Red Seal label. Stokowski and the Philadelphia Orchestra made Red Seal records exclusively from 1917 until 1940. And Koussevitzky made ground-breaking recordings with the Boston Symphony.

Most labels had an equivalent to Nipper and the Red Seals. Edison Bell buyers in Britain had a choice of Velvet Face, Winner or Bell discs to choose from. Velvet Face catered for the 'discriminating' music fan, manufactured from a 'special material' that ensured 'for them a silent surface. No foreign, harsh, or grating sounds are emitted from the V. F. Record'. These were reassuringly expensive at 3s 6d for the 10-inch, and 5s 6d for the 12-inch. Beneath these sat the populist Winners, where 2s 6d got you the 'ditties from the halls, the up-to-date dances, the breeziest band numbers, and other favourite melodies'. Finally, the 5-inch 'Bells' cost 1s 3d, and were designed for children, 'though they are not altogether without interest for the adult'.

While it's simpler to think in terms of the 'Big Three', in fact there was a complex patchwork of smaller labels across the music industry. Some survived, others faded into obscurity, some banded together to form mid-sized labels, others were swallowed up to become specialist series or imprints from within larger brands. Odeon Records, for example, started life in Berlin in 1903. It became part of the Carl Lindström Company – which in turn owned the likes of Beka, Parlophone and Homophon – and was eventually acquired by Columbia in the mid-1920s. Odeon started making lateral-cut 10-inch records of blue shellac in the United States, before being rapped on the knuckles for patent infringement and moving production to Europe, also releasing thousands of discs of music recorded in India. A German employee of

Lindström named Otto Heineman was trapped in America by the outbreak of the First World War, and eventually started out on his own, forming the beloved Okeh label. It started life issuing Odeon-made recordings in the United States, before being swallowed by Columbia in 1918. Meanwhile, when Indiana's Gennett Record Company was sued for making lateral-cut records, it successfully defended the argument that the lateral-cut method was now in the public domain, which suddenly created more elbow room for independents.

As labels squabbled over classical composers and opera stars for their high-class catalogues, they also catered for the latest dance crazes and newest sounds, such as ragtime, polka and tango, to blues, jazz and western swing. Black composer W. C. Handy's first popular success, 'Memphis Blues', was recorded by the Victor Military Band in 1914, and what's often described as the first proper jazz record, 'Livery Stable Blues', was recorded by the all-white Original Dixieland 'Jass' Band at Victor's 12th-floor New York studio in 1917. In 1920 Mamie Smith recorded 'Crazy Blues' for the Okeh label, vaudeville performer Ma Rainey began cutting songs for Paramount from the mid-1920s, and in 1923 Bessie Smith's 'Down Hearted Blues' sold 750,000 copies for Columbia.

'Rhapsody in Blue' gives us a spellbinding case study into what happened next. You see, in some respects, this 78 represents what was achievable at the very cutting edge of acoustic cutting. Just three years later, a largely similar group of musicians would record it again, using some newfangled techniques learnt from that smug, showy upstart: commercial radio.

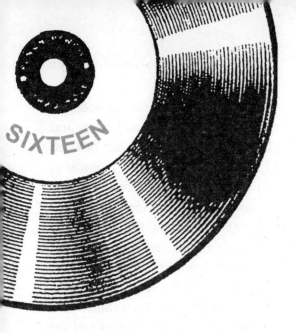

SIXTEEN

The Electrical Era

> 'Right here is the menace in machine-made music! ... The child becomes indifferent to practice, for when music can be heard in the homes without the labour of study and close application, and without the slow process of acquiring a technique, it will be simply a question of time when the amateur disappears entirely, and with him a host of vocal and instrumental teachers, who will be without field or calling.'
>
> John Philip Sousa, 1906

Radio represented a serious challenge to the phonograph. Some predicted the death of records in a gloomy narrative parallel to the demise of printed matter in the era of digital books. As bandleader Paul Whiteman wrote in 1948, 'mourners sang their dirge too soon'. In the end, radio proved to be a saviour, dragging the format from its shellac torpor and propelling it forward into the rock 'n' roll era, dressed in some fancy new vinyl clothes.

Commercial radio gets going in the 1920s. A quick evolutionary tour of radio begins with German physicist Heinrich Hertz, who was researching the transmission and reception of signals, and Canadian-American inventor Reginald Fessenden, who demonstrated voice transmission in 1900. In December the following year Guglielmo Marconi made his famous transatlantic transmission from Cornwall, England, to Newfoundland, Canada. Lee de Forest invented a three-element thermionic valve, which he called an Audion. Though initially intended as a radio detector, the Audion was a step towards practical electronic amplifiers. He helmed a series of experimental firsts – the broadcast of music from his laboratory in 1907, a spoken-word test of what the *New York Times* called 'wireless phones' in 1909, and live broadcasts from the Met the following year (which basically confirmed there was still work to be done).

In 1920, RCA's David Sarnoff proposed, in a 28-page memo, the 'Sales of Radio Music Box for Entertainment Purposes', which led to a shoring-up and cross-licensing of patents, putting the company in a position to dominate the nascent broadcasting industry. In the same year, a licence to broadcast for one hour a day was granted to KDKA Pittsburgh. Within two years the United States had around 500 licensed radio stations.

Throughout the first two decades of the twentieth century, phonograph engineers and cutters had got a lot better at their jobs. Mid-frequency range notes were fine, but very low or high frequencies still refused to behave. Early radio was plagued by oscillating whistles and interference, yet even with these drawbacks, it was far more satisfactory than the gramophone. You paid for it once, and that was it. The modest machine did away with pesky needles, records and all that brittle, rasping shellac, casually piping its music direct from the ether into your front room. Not only that, it sounded better. The best mechanical gramophone reproducers of the day could achieve excellent sound but were fundamentally constrained by the mechanics of acoustic recording and reproducing. The microphones used in radio, on the other hand, converted sound waves into electrical current, current that could then be amplified. A live band broadcast via a radio was in a different league. And radio's leg up to the phonograph industry took the form of a delegation from Western Electric, who approached Columbia and Victor in 1924, offering them use of a whole raft of gizmos that would make their lives a lot easier.

The first-known electrical recording on disc actually took place several years before, when four telephone microphones were placed inside London's Westminster Abbey to record the funeral service of the Unknown Warrior in November 1920. Designed by Lionel Guest and Horace Merriman, the carbon microphones were connected to a moving coil recording head. This was a very early prototype, and you can listen to the results – a recording of 'Abide With Me' from that service is fairly easy to track down. The first thing you'll notice is that the sound is awful, indeed noticeably inferior to acoustic records of the day. Nonetheless, it was a step. Bandleader Paul Whiteman writes about meeting the inventors in

Britain in 1923, investing money to bring their ideas to America, but without success. Then Orlando Marsh put out several electrically recorded sides via his own Autograph Records. This process used a microphone recording system to capture organ music – all but impossible before that date – putting out discs with Jesse Crawford playing a Wurlitzer pipe organ at the Chicago Theatre, alongside some recordings by Jelly Roll Morton.

While Guest, Merriman and Marsh were all technically first, their experiments remain something of a footnote in the history of electrically recorded music, partly because the sound wasn't markedly better than non-electric records of the day, and partly because they were up against a giant.

Bell Laboratories and Western Electric had enormous amounts of money and expertise at their disposal. They had been working on technologies for long-distance telephone transmission to span America, which would provide some of the building blocks for the new recording system. The first piece of the puzzle was the condenser microphone developed by electrical engineer Edward Christopher Wente. The old carbon mic was strong but not strong enough to cross the States. A condenser microphone, connected to an amp, could.

Around 1920, two teams at Bell Laboratories were created to pursue electrical recording: Joseph P. Maxfield and Henry C. Harrison oversaw phonograph recording, another team was developing a sound system for cinemas. Maxfield and Harrison used the condenser microphone, linked to a valve amplifier, with the amplified output voltage driving a moving magnet cutting head to cut the sound into a wax master disc. Or to put it another way, the

THE ELECTRICAL ERA

music would create a varying electrical output via the amplifier, which would go to the cutting stylus via the 'moving magnet' and inscribe the music wave form into the wax master. If the electrical signal exceeded certain limits, the cutting stylus tended to become overexcited and get stuck, so they added a tiny spring to limit movement. They also added an ingenious and hard-won component, the 'rubber line'. This was a rubber damping rod inside the cutting arm, which essentially worked as a kind of diplomat, soothing tensions between the bolshy electrical signal and the more relaxed mechanical cutter. For engineers, this rubber line would become one of their special areas of expertise – a component that required attention before each session, careful adjustment depending on the type of music they were recording. Together the new microphone, amp and electromechanical cutting gizmo become known as the 'Westrex' system.

It was time to show what they had to Victor and Columbia. They sent out 12-inch wax masters recorded from live radio broadcasts (they had been using radio to make test records throughout the design phase). We know from recording engineer Harry O. Sooy's memoirs that Victor received a batch of six recorded from a broadcast from New York City station WEAF. The actual recording equipment was made available in the latter part of 1924 for both Victor and Columbia to evaluate. To use the system, they would each have to make an up-front payment of $50,000, plus a variable royalty per record of around $0.01 per disc, but with a minimum annual royalty of $25,000. Victor hesitated, probably because of this high cost, as their profits had taken a serious hit in the wake of commercial radio.

An important figure at this juncture was Louis Sterling, who had arrived in London with a phonograph and £5 back in 1903. He worked for Gramophone, then British Zonophone, before starting his own record company, which was eventually bought by Columbia, from when he headed Columbia's new British subsidiary. On Christmas Eve 1924 he heard some of the test recordings (possibly through his sometime partner and our old 'Casey' stalwart Russell Hunting) and immediately realised this was a game changer.

Sterling tried to negotiate with Western Electric but found out that they weren't open to offering their licence overseas yet. Right, he thought, if they're only dealing with US firms, I'd better get one. American Columbia had been in trouble for some time and was carrying around a lot of debt. Sterling arranged for his Columbia to purchase the outstanding shares of the American Columbia, giving him the US credentials he needed to bag the Western Electric licence. Victor soon followed suit, and its half-ownership of Gramophone would eventually give the British company access to the new system too. By March 1925, the Electrical Era of Sound Recording had begun.

Early steps were tentative. Columbia's first electric cut was with the 'Whispering Pianist' Art Gilliam, while Victor tested the water with French virtuoso Alfred Cortot. Victor's electric equipment had been installed in their studios in Camden, New Jersey. The first few weeks were remembered by engineer Harry Sooy: 'We kept constantly on the go with this electrical recording ... Vocal solos, instrumental solos, vocal duets, symphony orchestras, dance orchestras and a mixed chorus of thirty-six voices ...'

The first Victor disc to hit the streets was 'A Miniature Concert' by 'eight popular Victor artists'. However, on 19 April 1925 Leopold Stokowski and the Philadelphia Orchestra made the world's first orchestral recording, *Danse Macabre* by Camille Saint-Saëns. Like 'Rhapsody in Blue', this forms something of a test-piece for our story, as Stokowski had attempted an acoustic recording of the same just the year before, which hadn't worked out very well.

The electrical system was able to record instruments at the low end of the frequency spectrum, below 200Hz. This meant that instruments such as the double bass, which in the old days had been replaced by tubas or bass clarinets, could now be recorded. The system could handle bass drums and other percussive effects too. Microphones were slowly setting the business free. Mics could be placed strategically around studios, recording different voices and instruments, set to different levels to bring everything together. No longer would piano players be balanced on platforms, or orchestras bunched around receiving horns. Players could play at a 'normal' volume, they could play with emotion, rather than just as loud as possible. Instruments that couldn't translate to disc suddenly could, and voices that wouldn't have impacted the grooves acoustically could be amplified and captured. Still, there were plenty of new problems to iron out: 'Empress of the Blues' Bessie Smith proved too powerful for the new technology and had to be recorded in a tent erected in the studio.

Records had suddenly become louder, and with a frequency range that had extended in both directions, yet many buyers wouldn't have necessarily spotted much of a difference to begin with. The advances in recording didn't immediately show themselves in the grooves, and when they did, most home set-ups simply didn't have the hardware to reproduce the new sounds with any degree of clarity. Tracking across the busier grooves, for example, was a problem, and many gramophone players simply screeched in protest.

The first version of the aforementioned 'Rhapsody in Blue' was recorded in the final throes of the acoustic era, in June 1924. The updated, electrically recorded version appeared in April 1927. Both were issued by Victor, on two-sided 78 shellac discs, and you can go listen to them right now and hear for yourself. Even though the three-year gap had resulted in stylistic changes in tempo, dynamics and arrangement, the huge step forwards in terms of pure sound is there to hear. Another example I strongly recommend you track down is Columbia disc 9084, recorded at the Metropolitan Opera House in New York on 31 March 1925. It's a joyful recording of 850 members of the Associated Glee Clubs of America singing 'John Peel', while the flip has the audience joining in on 'Adeste Fideles'. Go listen, then perhaps compare it to that historic live recording from the Crystal Palace in 1888. The groove had come a long way.

The marketing departments from both Victor and Columbia were concerned that if they made too much noise about their new method, it might devalue all their old stuff, leading buyers to write off the vast back catalogue of acoustic recordings that were still the firms' lifeblood. For that reason, both labels went through a period of experimentation, putting out discs that had been recorded electrically, but with very little fanfare. The average record buyer would not have noticed that Victor's *Danse Macabre* came from a new electrical process. Dealers were encouraged to use it as their demonstration disc on any record players, but the only physical hint that anything was different was a tiny 'V. E.' in an oval engraved between the run-out groove and the label. Columbia eventually adopted the name 'Vivatonal' and the words 'Electrical Process' in a lightning-bolt style font. And in October 1926, more than a year after the new recording method had begun, Victor gave its labels a fresh new look, with a 'Victor Orthophonic' brand, and 'V. E.' scroll logos to indicate electrical recording.

By the interwar period, Victor and Columbia were trading blows in the popular dance market. Victor had Paul Whiteman from 1920 to 1928, before he went over to Columbia. Through the swing era of the late 1930s, Benny Goodman recorded with Victor before defecting to Columbia. Glenn Miller started with Columbia, then went to Victor in 1938. Other names of the day, such as Guy Lombardo and Leo Reisman, flitted between the two.

Another flitter was Rudy Vallée, whose twangy crooning voice, aided by a handheld megaphone when singing live, was perfect for the new intimacy of the microphone. He sang with Columbia's low-priced labels Harmony, Velvet Tone and Diva, the scratchy Hit-of-the-Week label (whose records were laminated onto cardboard bases), then RCA's Bluebird and Victor, then

American Record Corporation's Perfect, Melotone, Conqueror and Romeo labels, before returning to RCA Victor.

Elsewhere, Victor's Ralph Peer held a series of recording sessions in and around Bristol, Tennessee, in 1927. These became known as the 'Big Bang' of country music and led to a series of landmark recordings of folk, blues, ragtime and gospel, as well as the commercial debuts of yodelling Jimmie Rodgers and the Carter Family (the same year Nashville's Barn Dance radio show became the 'Grand Ole Opry').

The microphone, radio and electrically recorded records were ushering in a new type of singer and a new type of performance. Suddenly, a vocalist could sing quietly, softly, with great emotion. Jack Kapp had an eye for business and an ear for talent. While scouting for Brunswick Records he recorded Jelly Roll Morton, had a hand in persuading Al Jolson to record 'Sonny Boy', and then met Bing Crosby in the lobby of the Roosevelt Hotel in Hollywood. Bing was dressed in a crew-neck sweater, giving off his weirdly powerful brand of magnetic, off-hand casualness. Kapp persuaded him to come to New York, where he cut his first records for Brunswick. Then, when Kapp founded the American arm of Decca, he promptly persuaded Bing Crosby to sign with him, following up with the likes of the Mills Brothers, Cab Calloway and Al Jolson. Kapp couldn't promise as much money as other labels at the time, but the stars had faith in his knack for knowing what would prove popular.

Kapp helped Crosby become one of the biggest stars of the day. He told him to cut out the whistling and 'boo-boo-booing' of his early records and persuaded him to branch out. Crosby would record anything that Kapp told him to record, from sacred songs to hillbilly, from Hawaiian to cowboy. American Decca, meanwhile, would also land The Boswell Sisters, 'Mrs Swing' Mildred Bailey and Louis Armstrong, alongside an extensive hillbilly catalogue.

Suddenly, there was electricity everywhere, amplification giving voice to brand-new sounds. Western swing, a joyful heady brew of country blues and Dixieland jazz, was filling up dance halls in California and across the South, and alongside solos from trombone and saxophone was the new Hawaiian sounds of lap-steel guitar. An example I strongly recommend you track down (if you haven't already) is the wonderfully named Milton Brown and His Musical Brownies. The Brownies had already been cutting western swing sides for some time when they were joined by a skilful player named Bob Dunn. Bob had rigged up some homespun amplification for his lap steel. This was the mid-1930s, and yet here is loud, lead guitar, with raw, primal snatches of distortion and feedback. At times he sounds like he can barely contain his amp, like an off-the-wall Marty McFly dropping audio bombs in the groove.

Bob Dunn's recording sessions with the Brownies in January 1935 mark the first appearance of an amplified electric guitar on record. Soon other players, such as jazz guitarist Eddie Durham, were amping up. Then, in 1939, Charlie Christian popularised the electric guitar as a member of the Benny Goodman Sextet and Orchestra. The legendary story of the Christian and

Goodman meeting goes like this: it was mid-August 1939. Producer John Hammond managed to secure Charlie some studio time with Goodman, but it didn't go very well. That night, Goodman and his orchestra were playing live at the Victor Hugo restaurant in Los Angeles. With more twisting of arms, Hammond managed to get Charlie into a spare spot on the stage. Goodman was displeased at this surprise, and called for them to play 'Rose Room', which he assumed Christian wouldn't know. Turned out 'Rose Room' was one of Christian's party pieces, a tune he had been playing for months. That version of 'Rose Room' would last 40 minutes, with about 20 solos, and by end of the night Christian was in the band. A couple of months later, he was featured in *Downbeat Magazine* with his famous call-to-arms, under the headline 'Guitarmen, Wake Up and Pluck! Wire for Sound: Let 'Em Hear You Play'.

The acoustic era has a certain magic. There's a romance, an artisan craftsmanship, a purity to sound waves being recorded via mechanical means. There's something beautiful about gathering physical objects of a certain shape, then arranging those objects in a certain way, which then gives them the ability to record sound and preserve it in the secretions of an insect. But now, microphones and electricity had given record engineers the means to capture sound with more precision and clarity than ever before. They had taken a can opener to frequency range, broadening the vistas of what was possible. The shellac groove could now be pushed even further. And radio wasn't done yet.

The Coming of the 33

> 'The Jazz Singer, that's what's the matter, The Jazz Singer!'
>
> R. F. Simpson, *Singin' in the Rain* (1952)

Shellac comes from the secretions of the lac insect. By the start of the Second World War, this had been the worldwide standard for record production for several decades. Solid shellac would have been far too fragile, so records started life as powdered resins and fillers added to a mix of ground-up insect goo, creating a thermoplastic substance that could be heated and moulded. The exact recipes were closely guarded secrets, and companies continually tweaked formulas to strike a balance between cost, quality and usability. That's not to say there weren't any non-shellac discs in this period. Of course there were: Filmophone records from the early 1930s, for example, were transparent, fully flexible celluloid in various gaudy colours. However, the vast majority were shellac, and some achieved less surface noise, others wore better, some seemed harder and heavier, others more brittle. The lac-producing bug was generally found in India, so before the country gained independence, Britain had a firm grip on the world supply.

Turning insect secretions into grooved audio is no walk in the park. The making of a shellac record was a long, technically demanding process, best illustrated by a brief fly-on-the-wall tour of a noisy American plant towards the end of the 1930s.

To make a record we need a master. This is the 1930s, so there's no convenient master tapes to work from. We must first record the music in wax.

First a thin layer of molten wax is poured onto a hot plate, a hot flame removes any bubbles or flaws. Once the wax is formed into a perfectly flat disc, it's slowly cooled. Then the inspected disc is passed through a slot to the recording room, placed on a turntable and the cutting stylus put in position. Masters would be cut using a jewel-cutting stylus. The cutting head in many respects is just like the pick-up of a record player in reverse, one difference being the feedscrew mechanism moves the head across the surface to cut the spiral groove whereas the pick-up on a record player just blindly follows the pre-cut channel of the record.

The sound engineer is ready, levels are set, the conductor and musicians poised, the music begins, the sounds pass through the microphones to the cutting head, to the stylus, which cuts the vibrations into the wax.

When finished, the soft disc is washed with nitrogen and put into a chamber with some gold. A 2,500V electrical current bombards the atoms of gold onto the wax. The gold-coloured disc is then placed into a solution of copper sulphate, and an electrical current transfers molecules of copper to the record. The metal disc is placed in a second bubbling, electrified solution. More electrolysis. Once the copper has taken a perfect impression of the record, the wax can be stripped away. This 'master matrix', as it's known, could in theory be used to make records right then and there, but it would not last long enough to make thousands. So another disc, this one called a 'mother matrix', is made first. This is given another bath, this time coating it with nickel, then it is bathed and coated in a fine film. It's back now for yet another copper bath, and now a mother matrix starts to build up on the face of the master. The resulting double disc is separated into mother and master – the master taken down to wherever the archives of masters are kept, while the mother is put in a nickel bath to give it a more durable surface. It's washed AGAIN, film coated AGAIN, and then it's put in a copper bath AGAIN, when the stamping matrix starts to build up.

Once this final plating process is complete, the mother matrix and stamper are separated. This last bit will happen several times over – they'll normally knock up several stampers, so lots of discs can be made at the same time. But wait, let's give that stamper another nickel plating, shall we? Then another coat of chromium? Yes, good idea. That should now be strong enough to last through many pressings. Just to make sure it's really tough, they fix the stamper to a rigid backing, then it's carefully centred, a hole punched in the middle, before a final wash and high-speed polish.

Now, most of the noisy action in the shellac business actually happened before this lengthy process. Making shellac suitable for records generally involved a lot of grinding, mixing, heating, rolling and slicing. The warm, soft shellac would then be rolled into a long, flat sheet before being cut into rectangular 'biscuits', each the right size to make one record.

Then came the pressing: two heated moulds, mounted face to face with a hinge at the rear. Two stampers, one for each side of the record, were fastened into place. The operator would sit beside a hot tray, on which were those rectangular biscuits of softened shellac. In went a label, then a soggy shellac biscuit, then a second label, then the press was closed. It was rapidly steam-heated, allowing the now-liquid shellac to flow into the tiny grooves, then instantly hardened with cold water, before being lifted out. Recently pressed records had their ragged edges trimmed by a circular cutting device, before being polished, inspected, then listened to. They were put into their sleeve, then into cardboard boxes and shipped off. Meanwhile, rejects ended up in bins, from where they were ground into powder, ready to be recycled.

There you have it, the shellac production line in the 1930s.

Now, this was all very well, but one way the radio would impact this laborious process was by speeding up that very first bit. In Britain, the radio touchpaper was lit in the summer of 1920, when the first live public broadcast took place from the factory of Marconi's Wireless Telegraph Company in Chelmsford, Essex, featuring the Australian soprano Nellie Melba. The BBC started out as a commercial company, founded in 1922 and working out of two rooms on the second floor of Magnet House, London, before the company dissolved into the non-commercial, crown-chartered British Broadcasting Corporation in 1926.

While there was no problem broadcasting live events, it became increasingly important to the BBC that its reporters and presenters could record and play back audio quickly. The contemporary method of producing records, with all that wax and electrolysis, was not suitable for a reporter on the run. A British inventor named Cecil Watts began to experiment with this problem in the early 1930s. By 1934 he had his first workable prototype, a method of cutting into a lacquer-coated aluminium disc.

Watts' lacquer blanks were essentially cut in the same way old wax blanks had been. However, these could be played back immediately. As the material was still much softer than a normal gramophone record, repeated listens would wear out the grooves more quickly. Nevertheless, the new system of 'instant discs' was particularly suited to recording on location. He developed cutting machines, too, including the 'model C disc-cutter', which could be carried in an ordinary car and used to record interviews. During the war this was replaced by a clockwork, portable, but still very heavy 'midget recorder', designed for reporters. The lacquer was in fact made of cellulose nitrate, the same substance as non-safety cinema film, which during cutting produced thin, highly flammable ribbons called swarf. All these machines tended to record at 78rpm.[26] There had been some use of 60rpm in the early days, and recordings were also made at 33⅓ rpm on 17-inch discs, using the same-sized grooves as a normal 78rpm disc, which could give up to 10 minutes per side.

If we return to our tour of the mid-century studio, the process of making the master is the same, but these lacquer discs are much easier to work with than their wax predecessors, and they sound much better. The 20mm layer of lacquer is soft enough to cut, but not so soft that the cut heals, and indeed will take several months to harden fully. The shape and substance of the carving stylus changed over the years – many engineers spending entire careers experimenting with what worked best – but they begin by using 'hot styli'. These have heated wires around the carving point, which momentarily soften the surface, offering less resistance, making a smoother cut with less distortion.

The cutting machine or lathe has to create a groove that spirals in perfect, even revolutions, cut with a uniform depth. Precision controls move the head that is slowly cutting into the blank disc. So that the operator can see what is going on, recording machines come with

---

[26] Record buyers of the era would also have had the chance to make records themselves. Shoppers could go to 'private' or 'instantaneous' recording booths in studios, shops or department stores. Again, here the lacquer disc was aluminium, coated with an extremely carefully controlled layer of black, smooth, shiny lacquer. Similarly, there were hundreds of companies offering private recording services – you could pay either to create a one-off acetate or to have a number of copies pressed.

microscopes that swing over the groove while the record is being cut, allowing the operator to check the depth, whether the grooves are far enough apart, and that there is sufficient undisturbed ground between the grooves – so the walls don't give way. There's also a suction device, to suck away the threads of cut material that the stylus throws off. Usually there's a load of test cuts, where blank grooves are made then examined, and played back to check noise levels.

Although proper talkies get going in the 1920s, attempts to fuse sound and film had been around since the start of moving pictures. A very early sound-film device was the Kinetophone, developed by Edison and William Dickson in the 1890s. This was a large cabinet, with a peephole in the top, and a listening tube. Essentially it showed moving pictures and combined them with appropriate but not perfectly synchronised audio. There followed the Cinemacrophonograph in 1899, the 'Phono-Cinéma-Théâtre' the following year, then Léon Gaumont's sound-on-disc Chronophone. In 1914, Finnish inventor Eric Tigerstedt demonstrated his experimental short *Word and Picture*, and this was just around the same time that Edison was back with a new projected version of the Kinetophone.

The primary difficulty was synchronisation. Yes, talking pictures that used shellac discs as the source for their audio were at least using a nice, familiar technology. The problem was film and shellac worked on wildly different hardware, devices that just did not play well together. As a result, many inventors had started experimenting with recording sound direct on film to do away with the problem. Lee de Forest, the inventor of the Audion tube, filed patents for an early sound-on-film process called Phonofilm. This method recorded electrical waveforms from a microphone directly onto the film, which could then be translated back into sound waves when the movie was projected. As they were already locked together, it did away with any complicated synchronisation. De Forest made a number of demonstration movies to interest major studios in Phonofilm, which include rare recordings of vaudeville and music hall acts, alongside film of de Forest himself explaining the system in 1922. These were the first publicly shown talkies. While sound-on-film would triumph over sound-on-disc in the end, for now the audio quality was just not up to the job and couldn't match the fidelity coming from Bell Laboratories.

Bell Laboratories/Western Electric were developing the Vitaphone, the first commercially successful soundtrack system. The business was established at Bell Laboratories in New York City, then acquired by Warner in April 1925. The first Vitaphone picture was silent film *Don Juan*, which was retrofitted with a symphonic musical score and some sound effects and shown alongside a 'The Voice from the Screen' short, where a Bell Laboratories bod explained the Vitaphone recording system, while also filming popular vaudevillian duo Witt and Berg.

The practical upshot of all this for the record technology was that the Vitaphone recording discs were specifically engineered to contain at least 11 minutes of sound. This wasn't picked at random – this corresponded to the length of one reel of movie film, so the two formats could be locked together. The origin, then, of

THE COMING OF THE 33

33⅓ rpm speed is simply that it was the speed that allowed these 16-inch Vitaphone discs to last long enough for one reel. They were played from the inside out, starting closest to the record's label. And the first proper talkie, although it's rather short on dialogue, was *The Jazz Singer*, starring Al Jolson.

In terms of the groove, there was no massive difference between these discs and ordinary 78s. The record-making process was the same – a lathe cutting audio spiral into the acetate-coated discs served as the master. They needed to be able to check that a take had worked immediately but, as even the lightest playback would cause damage to the wax master, it became standard practice to use two recorders simultaneously, so one wax could be played back in the room, the second sent for pressing if it was deemed to be a wrap. (Even when the sound-on-film system usurped the sound-on-disc, studio playback machines were used so they could check the audio immediately.) Meanwhile, Vitaphone-equipped cinemas, alongside the latest amplifiers and loudspeakers, had projectors that were mechanically interlocked with turntables. The projectionist would load the film, place the disc, then carefully position the stylus at a precise marked point on the surface.

Radio, movies and records weren't eating each other but instead building a symbiotic relationship. Similar format 'program transcription discs' were being used, which allowed broadcasters to make radio content that they could then sell and distribute to smaller local stations. RCA began making transcription discs of new 'Victrolac', a lightweight, flexible and less abrasive vinyl-based compound, although it did wear more quickly and wasn't free of surface noise. Their next move was to offer it for home use.

In September 1931, a group of musical celebrities and journalists were invited to the Plaza Hotel in New York. This was to be the first public demonstration of a new type of record, which was apparently capable of reproducing an entire symphony lasting a full half-hour. Led by RCA Victor president Edward Schumaker and conductor Leopold Stokowski, the select group was played Beethoven's Fifth Symphony, performed by the Philadelphia Orchestra. While Victor was in the habit of recording the Philadelphia Orchestra at their Camden studio, this disc (Victor L-7001) was recorded at the Academy of Music in Philadelphia. Despite the exciting promise of this new long-format disc, this first record was sold under the slightly utilitarian name of Program Transcription records. The grooves were essentially the same as standard 78s, but more closely spaced, and required a special 'Chromium Orange' chrome-plated steel needle. The 10-inch discs were shellac and contained popular and light classical music. The 12-inch discs were for more heavyweight fare and were pressed in new Victrolac.

This was a bad time to be launching new formats. The stock market crash of 1929 had decimated the phonographic industry, already wounded by radio. Record sales that had hit 110 million in 1922 would drop to just 6 million by 1932. The manufacture of record players had also ground to a halt. Into this environment RCA wanted buyers to not only buy new, longer, more expensive records but also invest in new hardware to play them. Two-speed turntables, such as their 1931 Radiola model RAE-26, which boasted the ability to play radio, 78 and 33⅓, were expensive, high-end items. Yet many of these RCA proto-long players weren't new recordings at all, just old 78s dubbed into the new speed, which wouldn't have pleased the very audiophiles they were hoping would take the plunge. Not only that, but the heavy pick-ups of the day tended to chisel through the Victrolac grooves with alarming speed, causing no end of complaints. They soldiered on, producing caches of Program Transcription discs in 1932 and 1933, including one-off titles for music halls and funeral homes, the latter labelled 'Sacred Music for Funeral Parlours'.

Edward Wallerstein, who joined RCA in 1933, couldn't wait to can the whole expensive enterprise. This chastening experience with RCA's failed LPs led him to have a dim view of the concept of long-players, and when he joined CBS, he become a significant point of resistance for Dr Peter Goldmark. However, we should all thank him for holding this line of resistance. As we're about to hear, he made life very difficult for Goldmark and his team, meaning they had to work harder, strive for better, to reach for perfection.

Discs at that time were sounding increasingly fabulous, but most people's home set-ups weren't yet able to manage the new amplitude. Victor launched a new line of 'Orthophonic' gramophones, and Edison launched a new heavier reproducer, with a spring-loaded stylus bar, which was available either as a new gizmo to breathe new life into old players, or came with the new line of

'Edisonic' machines, set free of rumble or hiss. The purely acoustic machine was on the way out, becoming the kind of relic your grandparents might own. Electric motors had been around since the early days of phonographs, but by now makers were replacing the acoustic sound box with an electromagnetic pick-up head. This meant that the needle, instead of vibrating a diaphragm, generated electricity. This current was fed to the valve amp and this to the loudspeaker. As electrical signals could also be fed to the amp from radio, the period produced the first hybrids – radiograms – where you could toggle between radio or groove with the flick of a selecting switch.

The Brunswick Panatrope arrived in 1925 and was the first all-electric record player. It was followed by an avalanche of clones, copycats and competitors. Before long, the first automatic record-changing decks appeared – made by Victor in the US, and a few years later by Garrard in the UK. Indeed, Garrard even produced a turntable that could play both sides of a disc without the need to turn it over. Edison, too, bows out of this story with a triumphant final flourish in the form of three new radio-phonographs, the C1, C2 and C3, which came with an ingenious pick-up designed by Charles Edison that could play both needle-cut and phono-cut 78s.

Throughout the 1930s and early 1940s, there was an increasing interest in the pursuit of high fidelity, pushing shellac to its limit. In April 1931, Leopold Stokowski and the Philadelphia Orchestra used a vertical-cut recorder equipped with a new moving coil pick-up with sapphire stylus, developed at Bell Labs by Arthur C. Keller to improve the dynamic range of cellulose acetate discs. When Stokowski listened to the record of 'Roman Carnival' by Berlioz, he said it was the finest he had ever heard.

Audio doyenne and long-time *Gramophone* scribe John Borwick described how engineers and cutters of the time would often tailor equalisation curve – limiting bass response to keep groove width to about 100-per-inch and boosting the treble. The result, in theory, was that this would translate well once the records were spinning in the living room, as home sets tended to restore bass and bring down the treble, so reducing surface noise. However, the actual recording curve tended to vary between labels, and even between engineers within labels. This meant that two records might sound very different at home – some might be overly boomy, others overly bright. Enthusiasts began making their own equaliser circuits – ones they could switch depending on the maker of the record, one labelled HMV, another labelled Columbia and so on. This homespun trend did not go unnoticed, and record companies began investing more time and money into ways of improving playback.

A hugely influential company in the expanding field of home hi-fi would be Decca. The company's lineage goes back to a portable gramophone called the 'Decca Dulcephone'. This was patented in 1914 by musical instrument makers Barnett Samuel, the name taken from simply merging the words Mecca with Dulcephone. The company would eventually be renamed the Decca Gramophone Co, but it really took off once it was taken over by an entrepreneurial stockbroker named Edward Lewis, who immediately started doing what he had been urging the Decca management to do for ages – make records.

In 1929, Decca ran studios at the Chenil Galleries on London's King's Road. Here, performances were captured by a single microphone, concealed from the musicians by a screen showing rural scenes. To begin with, in the wake of the crash of 1929, Lewis focused on trying to build up popular artists with mass appeal, while at the same time attempting to reduce prices, to undercut his main competitors, HMV and Columbia.

Decca's emergence as a major classical label was slow and steady but was partially built around names such as the great baritone Roy Henderson and his pupil Kathleen Ferrier. Their popular roster was fleshed out by taking over the bankrupt UK branch of Brunswick Records, which added Bing Crosby and Al Jolson to the books, as well as the Mills Brothers, Cab Calloway and The Boswell Sisters, alongside Decca's homespun hitmakers George Formby and the Band of the Grenadier Guards. A deal with German company Polyphonwerk gave Decca a sizeable classical catalogue, too. Net profits for 1937 were £273, the following year £15,080, the year after that £32,321. By the time Decca was celebrating its 10th birthday, it had studios in West Hampstead, a press plant in New Malden, and offices near The Oval.

Arthur Haddy started out as an apprentice making radio equipment at Western Electric Company. When he became engaged to the daughter of popular singer Harry Fay, he witnessed one of his father-in-law's recording sessions, at Crystalate. He was astounded at how primitive the equipment was. He joked that he could build something better on his kitchen table. Although only a passing comment, six months later the MD challenged him to follow through on his boast. The resulting amplifier and cutting head got him a job.

It was at Crystalate that he first began working with a young Kenneth Wilkinson, who would also become a celebrated audio engineer. During the 1930s, Haddy and Wilkinson gradually developed improved cutting heads and dynamic pick-ups, with the aim of widening the usable frequency range of his records. In the end Haddy focused on moving-coil pick-ups, producing a series of records that, though having not much improved frequency range in the end, certainly sounded tip-top. Crystalate was taken over by Decca in 1937, and Wilkinson and Haddy both joined the ranks.

After 1939, Haddy was moved from commercial recording to war work. His task was to make recording equipment that could detect the sonic differences in the water movement around German and British submarine propellers, so that crew could be trained to tell the difference. This was right at the limits of what was possible; the differences were at the high end of the frequency range. Eventually Haddy managed to meet the requirement by doubling the response of his recording cutter head to 15,000Hz.

This technology would form the basis of Decca's 'FFRR' discs – Full Frequency Range Recordings, which were sold from 1944. Wilkinson later worked with Decca's classical producer John Culshaw, who was masterminding the first studio recording of Wagner's *Der Ring des Nibelungen*. This Herculean task began in 1958 with *Das Rheingold*, the shortest of the four operas.

Culshaw wasn't a fan of recording live opera. He felt it was merely the sound of a theatrical performance, and preferred an approach he called 'theatre of the mind', where production techniques were used to conjure up the action in the listener's imagination.

Decca's marketing department promoted Culshaw's recording technique under the name 'Sonic Stage'. To give an example: the *Das Rheingold* score calls for 18 anvils to be hammered during two brief interludes, an instruction that was never followed in opera houses. Culshaw wrote how opera lovers were 'fobbed off' either with some sort of electronic compromise or by the 'tinkling sound' of a few people beating metal bars together. What you didn't get, he complained, was the 'firm, frightening sound' of 18 anvils hit with rhythmical precision, which would become a 'deafening assault' on the nerves. For the studio recording, Culshaw arranged for 18 anvils to be hired and hammered. The reviews were ecstatic. *Gramophone* called it stupendous, surpassing anything done before. And to the great astonishment of the outside industry looking in, the set outsold Elvis Presley and Pat Boone.

The Long Players

'We held our debut at the Waldorf. It didn't make big waves.'

Dr Peter Goldmark, 1973

The Ageing Edison could be characterised as a meddling Luddite who didn't understand music. He had clung to the cylinder too long, stuck with hill and dale, and then dragged his feet over electrical recording. He had strong views about what constituted music as well, not caring for anything too highbrow or complicated. Perhaps because of his partial deafness, he liked a simple tune, sung by a single voice with scant accompaniment. He was like everyone's grandfather, muttering in the corner that music was much better in his day, tunes you could hum along to. In spite of

Edison's dim view on the merits of electrical recording, his son Charles and Walter H. Miller managed to introduce the new system in 1927, a little over two years later than everyone else.

Before we say goodbye to him for good, though, let's not forget that, while he was something of a pig-headed hindrance towards the end, he did create some of the finest cylinders ever produced, make some of the loudest and sweetest-sounding machines of the acoustic era, record arguably the finest phono-cut discs the world had ever seen, and, if that weren't enough, pioneered talking pictures, and made some long-playing records, a full two decades before Dr Peter Goldmark invented the LP.

The vinyl long player, when it finally arrived in the 1940s, would have about 250 grooves per inch. Edison's little-known long-play discs had 450 per inch. Produced in 10- and 12-inch formats, they revolved at 80rpm, giving 15 to 20 minutes' playing time per side, marketed as being able to provide dinner music 'from soup to nuts'. Sadly, the grooves were too fine, the discs too quiet and too fragile. They were issued just as electrical recordings, noticeably louder and clearer than their acoustic predecessors, were flooding the market. And rather than using the extended format to record longer pieces hitherto impossible, they chose to fill the hair-width grooves with dubbings of four-minute tracks already available as standard Edison discs. As a result, the Edison Long Play became another forgotten footnote in vinyl's origin story. However, it illustrates that producers of the time were pushing at the edges, always seeking to increase capacity, to get the most out of shellac. And let's also keep in mind that, while the

Edison long-play discs were not a success, they did work … as long as you treated them with a lot of love and a great deal of care.

So, aside from Decca's FFRR, the humble record had hardly changed for the average record buyer. Radio continued to improve its transmissions and receivers, yet shellac's shortcomings were much the same as before – emery powder rubbing up against the relatively heavy pick-ups, which averaged about 140g of downward pressure. And the main bugbear was still run-time. For the dance-hall DJ it was fine – a diverting foxtrot or novelty song could easily fit on one side – but for the classical buff things were a long way from ideal. The only attempts at shellac-based long play had failed. There were the 20-inch Neophones from 1904. Boston label Grey Gull Records had recorded and manufactured vertical-cut discs, which were the earliest records to contain more than one track – known as '2-in-1s' – as the smaller grooves gave about twice the playing time of standard 10-inch discs. There was the British pre-war label Musogram, which issued hill-and-dale 'Marathon' records, making a big noise about how they could fit the entire *Flying Dutchman* overture onto one double-sided 12-inch record. Finally, there were Edison's Long Plays and, of course, RCA's failed Victrolac discs.

It was into this environment that the LP was launched, at the Waldorf-Astoria Hotel in New York, in June 1948. It's unlikely there was much excitement to begin with – there had been plenty of events like this before, although this one did seem more carefully choreographed. An A&R man stood before the assembled hacks with two piles of records – one several feet high, the other up to around knee height. Then he told them that both piles represented exactly the same amount of music. That got a few writers shifting in their seats. They played a piece of classical on a 78, which was soon interrupted by the disc running out of space. Then they played the same piece on the new LP, which not only sounded better but played it all the way through.

No one hated the phonograph more than Peter Goldmark. He had always hated it. It seemed to him to violate and spoil what it set out to recreate. The machine had learnt to talk but hadn't yet learnt to sing. This feeling of dislike was only cemented by the rapturous reception others gave the phonograph. While in England during the war, one of his Army friends kept playing the same song over and over, in apparent delight. Yet to Goldmark it was an appalling cacophony of tinny, scratchy clicks, mixed with music in a 'discordant mess'. He had to leave the room.

Shortly after VJ Day, Goldmark was visiting friends in Westport, Connecticut. After dinner they played a new recording of Brahms' *Second Piano Concerto* by Vladimir Horowitz, Arturo Toscanini conducting. In the midst of the record, there was silence, strange noises, then a click, then the disc was changed. A few minutes later it happened again, then again. Goldmark counted 12 sides for the four movements, making 11 interruptions, of which eight were unplanned by Brahms. In his biography, he compared the jarring experience to receiving a phone call while having sex.

Goldmark felt there *must* be a better way. That night he took out a pencil and a pad, and began making calculations about the widths of groove, the wavelength of sound on record, the speed of revolutions. Changing the speed was a simple enough operation, but changing the width of a groove would present more of a challenge. Still, the idea of a record that could do the job better kept him up that night. For further research, he went on a guided tour of the company's record plant in Bridgeport and its broadcast operations. Alongside the maelstrom of activity that went into the shellac pressing business, he became aware of the 16-inch, 33⅓ rpm lacquer transcription discs that were recording CBS broadcasts. This seemed an excellent starting point from which to build a new long player.

Goldmark was given some time with the then head of Columbia Records, Edward Wallerstein, who listened to his idea for about three minutes, then put his arm around him and suggested that he should go back to doing television work. Wallerstein's reaction isn't surprising. He had been at RCA when their 33⅓ rpm transcription discs had failed to sell and had considered it a triumph that he managed to pull the plug. To him, Goldmark was just some scientist guy – not only that, but a *televisual* scientist guy – saying: 'You know that idea that didn't work? I wanna do that.' Thankfully, Wallerstein's patronising reaction only angered Goldmark and made him more determined than ever to succeed.

He went back to investigate the old RCA transcription discs. As far as he could see, they had just slowed down the record and made it worse. He wanted to slow down the record and make it better. As something of an outsider looking in, it seemed to him that the whole recording business was a dinosaur. It had a strange, mechanised, old-fashioned culture, which had got it to where it was through trial and error, with very little will to improve. Things were working, so why fix them? Still, Goldmark felt that one area where he could make a difference was the disc material itself. He started to think about a smooth, hard material that could replace shellac.

Vinylite had been developed during the war. It was unbreakable and light but also cost twice as much to make as shellac. However, the more Goldmark thought about it, the more he realised that if you managed to squeeze more music onto a single disc, it would in the end reduce costs. He compared lengths of classical movements. The average classical piece seemed to take about 36 minutes, and the vast majority of all classical works could be fitted into around 45 minutes of playing time. That became his target.

It didn't take long to calculate what would need to happen to produce that much playtime. He settled on the 12-inch record, since it was already the standard, and determined the number of grooves per inch that would be required. Next came the tricky parts – finding the right pressure and material with which to cut the thinner groove and to preserve the gossamer-fine cuts, and then how to get the best sound from those grooves.

This is the thing. You might think: oh get on with it, make better records. And yes, on the surface, the process of making vinyl wasn't that different to making shellac. But Goldmark was proposing waging format war, upsetting a balance that didn't need to be upset. To make this work, he would have to look again, not only at the surface material and width of grooves, but at everything. He'd have to look at cutting styli, lathes, presses, pick-ups, turntables, amplifiers, and, he realised, even packaging. He was aiming to change not only the physical make-up of the record, but the culture of buying and consuming records. Still, he didn't quite grasp the magnitude of what they were attempting to begin with.

He took his plan to CBS executive Paul Kesten, mentioning that he had already been turned down. Kesten didn't seem bothered, and simply asked how much the engineer thought he'd need. When Goldmark said $100,000, Kesten replied, if you can do it, that's fine – we'll finance it.

Goldmark hired a shy but friendly Belgian recording engineer named Rene Snepvangers, who had been working at RCA, cutting NBC transcription discs. He didn't give him any detail at first, worried that if he didn't accept the job, he might go back and tell RCA what they were up to. When he did finally tell him, the Belgian immediately trotted out a long list of reasons why they should immediately stop trying.

More technicians joined the team as they started running tests. They began tinkering with sapphire styli of the day, to see if they were able to ground out the lacquer in a smooth, thinner, but continuous thread, rather than chipping, which would spoil the record. They continued to work on the problem, cutting grooves with a gradually perfected recording head, and playing back with a pick-up that could track the thinner grooves. Their first landmark was a record that played for 15 minutes on one side, then 22, then 25. One even managed half an hour.

They produced a vinyl prototype, choosing a rendition of Tchaikovsky's violin concerto. The first sounded poor – to Goldmark's ears, the violins sounded like flutes – but with each test cut, they got closer. Wallerstein, who was a constant threat to the entire project, wasn't impressed. 'Where's the fuzz?' he asked indignantly. The fuzz, it turned out, was the sound of violin bows, the scrape of the resin. He wanted to be able to hear the fuzz. It was good, he conceded, but not good enough. They might have made a longer record, but it wasn't better yet.

Back in the lab, Snepvangers suggested they make test records of gunshot, as this gave them a strong, recognisable sound with which they could compare their test cuts. That way they could try to pin down and iron out the problems in the recording process that were holding them back. Soon they were firing shots into heavy mattresses, playing back the audio, which, at first, sounded like someone dropping a baked potato.

They continued gunshot tests, using different components along the production line from mic, to cutter, to stylus. They concluded that the problem lay in the ribbon mic, which was affected by 'phase distortion'. This was a fairly common issue with microphones, caused by frequencies that went into the mic in a certain order failing to arrive at the end of the ribbon in the same order, and so ending up sounding garbled. They tracked down a German company that had just started making a new type of condenser mic that could avoid this problem. With the new mic installed, the difference was enormous. It sounded exactly like the gunshot and nothing like a baked potato.

The first music to be recorded in this way, using the condenser mic, was an informal trio of staff in a secret studio on the 10th floor. One engineer was on violin, a secretary played the piano and Goldmark brought in his cello. They set up the new mic, played some Bach and listened back to the first LP record. They immediately called in Wallerstein, who exclaimed, in a rare moment of levity, they'd got it!

The aftermath to this lovely moment was more hard slog. Almost immediately, Wallerstein saw only problems again. Great, you can record a trio, but what about an entire orchestra? And what about all the old 78 masters in the archive? How on earth were they going to transfer them?

This last problem was a serious one – a headache that Goldmark's team had, for now, avoided thinking about too much. They would spend the next weeks and months figuring out how to seamlessly splice together endless segments from the original shellac cuts in the company archive, which took a great deal of work. At the same time, company president Frank Stanton held a company-wide contest to come up with a name for the new discs. There were 25 suggestions, 25 rejections. Goldmark remarked that perhaps 'the LP' wasn't going to have a name after all.

Startled, Stanton said: 'What did you say?' And that's how it got its name.

CBS signed an agreement with the Philco Corporation, who began manufacturing the means to play the new discs – new players, and a cheaper attachment that could be fitted to traditional pick-ups. The original cohort of LPs included eighty-five 12-inch classical LPs, twenty-six 10-inch classics, eighteen 10-inch popular numbers and four 10-inch juvenile records. The first was ML 4001

Mendelssohn's *Concerto in E Minor* by Nathan Milstein on the violin, conducted by Bruno Walter. Other highlights included a 10-inch reissue of *The Voice of Frank Sinatra* (CL 6001), and the original juvenile record was *Nursery Songs by Gene Kelly* (JL 8001). They even experimented with a 33⅓ rpm 7-inch as part of this new long-playing suite. But in the end, Goldmark decided not to show this prototype to the bosses, worrying that a smaller, slow record would only confuse them.

At first, production did not run smooth. In theory, there wasn't much difference in the art of pressing shellac or vinyl. Microgroove pressing was a more delicate operation, so the process per LP disc was slower. But if you think of it in terms of the total playing time that was produced, it was quicker. A new LP was made with one set of stampers. The same run-time on 78 records would have required 12 or more stampers. Nevertheless, impurities in the vinyl were causing ticks, jumps and surface noise.

They returned to the Bridgeport plant, where by now a section had been given over to the new vinyl press. However, it was near all the old shellac business, which caused no end of dust. Goldmark pointed this out and the workers dutifully gave the place a clean, causing even more dust. It took time to change this working culture, to drum into people that these new LPs were tougher in some respects, they wouldn't shatter or snap, but were much more delicate in others. Finished discs were often misunderstood and mistreated. First-generation users might just give the surface a quick rub with their sleeve before placing it on the turntable, as that's what they had always done with shellac, even though this always left behind more dust than it cleared.

As they continued to amass a generous launch catalogue, the heads decided to invite the competition to see their new disc in action. On the day of this summit meeting, the RCA head David Sarnoff filed into a Columbia boardroom, flanked by a group of his engineers. The atmosphere was tense but cordial. Goldmark was nervous. The impeccably dressed CBS chief William Paley indicated what was about to happen. Goldmark would first play them a segment from an ordinary 78, then would follow it with their latest invention. Goldmark remembered that Sarnoff seemed to stiffen. He approached the turntable and played the 78 for about 15 seconds. Then he put the needle to the LP. Within the first few bars, Sarnoff was on his feet. They switched between the two discs again, by the end Sarnoff warmly congratulating them on the achievement. Then came Paley's proposal: they would delay telling the world about their new LP, and they would share their know-how with RCA and they could join forces. Sarnoff thanked him for the offer and they left.

Soon after the LP's launch, at an annual sales convention in Atlantic City sales manager Paul Southard wrote a speech that was exactly the length of *The Nutcracker Suite*, which was on one side of an LP. He began speaking just as the stylus was placed on the record, which continued playing softly in the background throughout. When the speech ended, and he lifted the stylus, the 'distributors went wild'.

Now, to tell the story, we've been listening in large part to Goldmark's version of events. In that version, many of the original ideas and impetus for the LP came from him. But let's keep in mind that, after the LP had caught on, Wallerstein too gave interviews about how the project came together, and what's notable about these is that Goldmark is barely mentioned. In Wallerstein's version of the story, *he* was the one who timed a load of classical pieces to come up with a target run-time for the new format. In his version, he wasn't even interested in listening to the records until they were over a certain length – continually knocking back approaches from the engineers, saying 'that's not long-play'. In his version, *he* plays the record for the RCA contingent, *he* comes up with the name LP. According to him, the reaction to the LP's launch at the Waldorf was fantastic (Goldmark remembered it being low-key).

Indeed, you get the impression Wallerstein is angry that Goldmark has become known as the LP's 'inventor'. No one person could be said to have invented the LP, he argues, as this was not an invention but a 'development'. He says that Peter was merely the supervisor, adding that 'he didn't actually do any of the work'. Wallerstein is full of praise for the contribution of Rene Snepvangers, who concentrated on the problem of developing the lightweight pick-up, and in particular Bill Bachman, for his work on the heated stylus, automatic variable pitch control, and most especially the variable pick-up.

In the end, many engineers and managers deserve credit for the breakthrough: Goldmark, for overseeing; Bill Bachman, as research director; as well as Bill Savory, Ike Rodman, Jim Hunter, Vin Liebler and Rene Snepvangers. They had approached the project to improve and revolutionise an old format. Looking back, Goldmark felt that more than anything, their work sparked a completely new industry – an obsession and insatiable demand for ever-higher fidelity. The disc would eventually change the way labels, musicians and consumers packaged, created and listened to music. At the time of launch, the very word 'album' referred to collections of discs. As LPs gained traction, 'album' came to mean a collection of songs or pieces within a single record. They had created a new standard unit of music making. LPs stopped being collections of singles and B-sides and second-string studio takes, but creative expressions through which bands and musicians began to conceive common themes and narratives.

Columbia was in a good position to exploit the new LP, as engineers were already using Ampex machines, and beginning to curtail direct disc-cutting. Knowing that they needed the new format to catch on, they made their hard-won technical know-how available to any who wished to use it, and it wasn't long until other companies began issuing LPs. In the States, the first were Vox, Cetra-Soria and Concert Hall, then Capitol, Mercury and Decca. Soon Columbia reduced the price of the Philco attachments from $29.95 to $9.95, and this loss leader resulted in more record sales, and more companies beginning to bring to market their own LP-playing equipment. Before long the 'Tombstone' – the first jacket design used on most of their early LPs – became a familiar display in record stores across America.

In the immediate aftermath of that tense meeting in the Columbia boardroom, however, Sarnoff was livid. A few days later, he phoned Paley to say RCA had decided not to go in with them. This may have been because he doubted their new product would be a success – indeed, there were plenty on both sides who doubted it would catch on – but it may also have been because they had something else up their sleeve. Something they were calling 'Madame X'.

NINETEEN

The Speed Wars

'Based on the above considerations, the speed of 45 is obviously superior to either 78.26 or 33⅓ revolutions per minute, and was therefore chosen for this system.'

RCA Review, June 1949

Before we continue, here's a quick reminder of some attributes of the lateral-cut groove. The depth remains constant, while the width of the groove varies depending on the strength of the signal, which in turn impacts capacity. Records produced with a single lateral modulation provide one channel of audio (we'll come to how they figured how to squeeze two channels into a single groove later) and it's important to remember that lower-frequency sounds take up more physical space than higher frequency. To offset this effect, engineers began reducing base frequencies, boosting higher frequencies, before cutting the disc. And while there are of course wide variations and tweaks, the 'RIAA curve' would become the most widely adopted equalisation standard from 1954 – songs engraved with the volume of low frequencies reduced, high frequencies boosted. This filter would then be corrected in home amplifiers, which would re-boost the bass, reduce the treble.

Anyway, some of those first-generation LPs sounded dull. The longer notes wavered and the louder passages sometimes appeared in faint pre-echoes ahead of time. And while some contemporary audiophiles took to the new format with gusto, others exercised caution. Like any good format war, it takes a while for buyers to open themselves up to the new way of doing things. There's that nagging feeling that you're being taken for a ride, being persuaded to re-buy things you've already bought. It's a leap of faith, and an acknowledgement that the days of your old collection are numbered. However, there were obvious and marked advantages. Record buyers could now listen to lengthy passages of music in their entirety, there was much less surface noise and the records didn't wear out so quickly. New hardware aside, they weren't too expensive either – a recording of Beethoven's *Fourth Symphony* on five 78 discs had cost $7.25. The same music on a single LP record cost $4.85.

The Waldorf Hotel launch's immediate impact was to kick RCA into action. After some months of ominous silence, they returned to the ring with their own microgroove product, this time in snazzy 7-inch form. Columbia may have had their Dodge Winnebago, but now RCA had a Porsche. What is sometimes forgotten, however, is that to begin with this was not marketed as a throwaway 'pop' vehicle but a shrunken version of the old 78 multi-disc albums, as batches of 7-inch album 'sets', which were to be played on a new type of player that boasted the 'world's fastest changer'.

Ads run in May 1949 announced RCA's new 45rpm format under the heading 'The remarkable background of Madame X'. The copy goes on to explain, in a lovely piece of advertising hokum, that Madame X was the code name that had been given to the disc during years of research and development. It talks about the long quest for 'tone perfection'. The result, it says, was a new, completely integrated record-playing system – not only brand-new records, but a brand-new kind of automatic player, that between them were the first to be 'entirely free of distortion'.

In choosing 7-inch and 45 speed, the RCA engineers were aiming to hit three parameters: a playing time of up to five and a half minutes per side, a terminal linear velocity of around 11½ inches per second and maximum grooves of 275 per inch. They also argued that as speed of a record was reduced, the volume of material required increases, meaning that 33⅓ rpm was *much* less efficient than their 45rpm disc.

Naturally, the idea of knocking up a brand-new, cutting-edge record changer wasn't a simple proposition. And they didn't want the changer to just do the job, they wanted it to excel. This wasn't to be a cross-format, multi-speed player either. It would only play RCA's 45s, and so they needed it to blow the competition out of the water.

The default quarter-inch central hole that all records had at the time was a problem. It was too small – the automated changing of a record couldn't happen quickly enough. By widening the central hole to about an inch and a half, and making a broader, more generous spindle, it allowed the new 45s to drop into place more quickly and easily. The wider spindle also gave the discs more support.

The new fat spindle was about 6 inches high and so could support a modest tower of up to 10 discs. Once one side had finished playing, the tonearm moved away, the next dropped down onto the platter, then it swung back and play would continue. It was theoretically possible for the music to continue for a full hour before anyone had to touch anything again.

RCA started by selling sets of four to six 45s, each providing about the same amount of music as one LP. In one extreme case, the complete recording of *Carmen* with Risë Stevens and conducted by Fritz Reiner was issued in a collection of sixteen 45rpm discs.

> The record is of non-breakable vinyl plastic, wafer-thin. Yet it plays as long as a conventional 12-inch record. The new RCA Victor automatic record changer holds up to 10 of the new records – 1 hour and 40 minutes of playing time – and can be attached to almost any radio, phonograph, or television combination.

Goldmark couldn't believe what RCA were trying. He attended live demonstrations, and to him this seemed a backward step. However fast a changer was, it was still changing the record. And as this was the very thing he and his team had sought to cut out, the fact that the competitors were keeping it in didn't make sense. Surely no one would choose this weird multi-pack 45-speed throwback over their new 33⅓? However, they were up against an absolutely enormous advertising budget, and RCA went big. The upshot of these opening salvos in what the popular press would call the 'War of the Speeds' was a short but damaging blip where, in sheer confusion, many people stopped buying either.

Viewed from a certain direction, RCA Victor's approach wasn't as weird as it sounds. When they first started work on Madame X, they were still only competing with the 78. And even though the 33⅓ rpm LP was new in town, there were still millions of record buyers who hadn't yet crossed the tracks to the microgroove. Many were buying and listening to those hefty 78 multi-disc albums. To them, the idea of a set of records, much clearer, stronger, smaller and lighter, that did the same thing but quicker, must have appealed.

Still, it would be an oversimplification to suggest that RCA only designed the 45 with these multi-disc album collections in mind. Popular *and* commercial considerations shaped the 7-inch. An issue of *RCA Review* from 1949 describes the work that went into the design of Madame X, which included a comprehensive study of run-times within the Victor catalogue. This focused on the 'Music America Loves Best' series, a group of popular bestsellers. They also studied 'music units' of classical pieces from the Red Seal label, defining 'units' as a selection or a part of a work, such as a movement of a symphony, that was written to be played without a break. This study concluded that most popular songs and a high percentage of those classical units were less than five minutes long. Then, in a passing dig at the competition, they concluded 'it would appear from this that undue weight may, at times, have been given to the importance of long-playing time'.

The very first 45rpm record manufactured was 'PeeWee the Piccolo' (RCA Victor 47-0147), pressed on a 7-inch Madame X on 7 December 1948 at the Sherman Drive plant in Indianapolis. To begin with, the 45s were issued on coloured vinyl – each genre given its own colour. Pop was black, musicals blue, classical red, rhythm and blues orange, and so on. This was soon discontinued, all records becoming the standard black.

Columbia reacted by putting out their own 7-inch disc, set to revolve at 33⅓ rpm, and their advertising of the day often went with the slogan 'One Speed is all You Need'.

By the late 1940s, more record companies were jumping on the longer LP bandwagon, and when American Decca joined, it became something of a stampede, with every major label following suit, aside from RCA. Then finally, in January, they did, announcing a new series of classical long-playing discs.

All the while, RCA continued to spend millions of dollars advertising the 7-inch. While it failed to outperform the LP in key areas, it did have advantages. And soon, with relentless advertising, it would become the favoured receptacle for popular music, completely replacing the 10-inch dance 78s. By 1954, sales of the 45 overtook the 78 for the first time, and soon the same record companies that had come over to 33⅓ were also going over to the 45, with Columbia following suit. Capitol became the first major label to support all three recording speeds.

There was bad blood over the format wars. Most of the vitriol seemed to be aimed at Victor RCA for introducing their 45 when they did. The music industry was still dealing with the aftermath of the long-running musicians' strike. Many dealers were tearing their hair out at having to grapple with music being available across three formats. And the buying public were equally miffed at the confusing hardware choices. It killed off shellac in a more brutal way, meaning many old discs that might have had a chance to be transferred to vinyl had the end come more slowly just ... disappeared. By the end of the unpleasantness, the formats were no longer fighting each other but coexisting. The 78s hung on long enough to sport singles by Elvis Presley, but they were doomed. Shellac finally shuffled off. Noisily.

While we're here, let's just check in with a bit of popular culture, shall we? We've had our fill of scientists and suits, opera buffs and classical snobs, and frankly, I'm sick of them. Remember this: 'PeeWee the Piccolo' was pressed in December 1948. In just over a year, a man called Sam Phillips will open the Memphis Recording Studio. A few months after that, he'll produce 'Rocket 88' by Jackie Brenston and His Delta Cats – in fact, Ike Turner and His Kings of Rhythm. Following the success of 'Rocket 88', he will found Sun Records. And pretty soon after that, an 18-year-old named Elvis will wander in off the street and cut an acetate of 'My Happiness'. And not too long after that, he'll record 'That's All Right'.

The LP gradually cemented the album as a unit of music, while the lightweight, inexpensive 7-inch held the three-minute pop single in its grasp, which remained the standard into the 1960s and beyond. Slow progressions in mastering techniques enabled longer songs; a famous turning point came when Columbia bowed to artistic pressure and cut the six-minute version of Bob Dylan's 'Like a Rolling Stone' into a 7-inch single. The impact of all these shenanigans for the coming generations of record buyers was that records either came with a large hole, which had to be played on standard players using an adapter or, in some regions, 7 inches would arrive with a default smaller hole, within a 'knock out' section that could be removed for larger hubs.

Shows, Soundtracks and Sleeves

'You know you haven't stopped talking since I came here? You must have been vaccinated with a phonograph needle.'

Groucho Marx

Towards the end of February 1949, there was a great deal of nervous excitement as a group of A&R men were gathered at Richard Rodgers' apartment. They were there to hear songs from his latest collaboration with Oscar Hammerstein: the musical *South Pacific*.

The LP really began taking off with musicals. In a sense, musicals would do for the LP what Bing Crosby had done for electric recording. The original Broadway production of the first Rodgers and Hammerstein musical, *Oklahoma!*, opened on 31 March 1943 and was an absolute smash hit, running for 2,212 performances. For a while, the buying public couldn't get their hands on any of the songs, because musicians were on strike over royalty payments. This meant that union musicians could play live or perform on radio but they were not allowed to record. Decca's president Jack Kapp settled with the union in September 1943, and three weeks later dragged the *Oklahoma!* cast and orchestra into a recording studio. A streamlined *Oklahoma!* appeared in an album of six 10-inch 78s, selling more than a million copies, and soon they called the cast back to record the additional songs that had been left out. *Oklahoma! Volume Two* was released in May 1944.

This wasn't the first Broadway hit to be put on record. Back in 1933, Decca had produced hit collections of songs from *Blackbirds of 1928* and *Show Boat*, for example. But before *Oklahoma!*, songs from the latest musicals tended to be recorded by other popular singers of the day, with only light orchestration. Recording the original cast with full orchestration offered record buyers the chance to experience the musical as it had been performed, and this really caught the public's imagination. The next Rodgers and Hammerstein musical, *Carousel*, did well for Decca; the third, *Allegro*, was something of a disappointment for RCA. However, even a relatively poor-performing cast recording was offset by hits of individual songs.

Since the launch of the LP, Columbia's vice-president Goddard Lieberson had overseen the repackaging of their 78 recording of *Finian's Rainbow*. He scored a huge hit when he bought the recording rights to *Kiss Me, Kate* before it had even got to stage. The cast album, recorded on LP and released in 1949, sold more than 100,000 copies in its first month. And now he saw *South Pacific* as an incomparable opportunity to induce consumers still dragging their feet in the early exchanges of the format wars to embrace the LP for good.

There was already a lot of anticipation. The songs of *South Pacific* had been made available to all the major labels even before the cast had started rehearsing. Frank Sinatra and Perry Como recorded 'Some Enchanting Evening' for Columbia and RCA, Decca and Capitol put out six-song album sets and, just as the musical opened at the Majestic Theatre, Peggy Lee was recording 'I'm Gonna Wash that Man Right outta My Hair' in Los Angeles. Despite coming first, all these did was heighten interest in the forthcoming cast recording.

Lieberson took a firm grip over the project, not satisfied to simply put six songs on either side. He wanted to use the elbow room given by the new expanded format to help tell the story of the musical. This meant tweaking various elements, tailoring it for home listening. The main example of this comes from the end of the show. On stage, there were some obvious directions that gave the musical its romantic conclusion. Without actors to show what was going on, he felt the subtleties would be missed. So for the record, he took a section from the first act, to use as an on-record finale.

South Pacific was released on both LP and 78. It stayed at the number-one spot for 63 weeks, the longest run of its time, and an astounding achievement unmatched for a decade. *Variety* reported how Columbia's new 'pop' items, *Kiss Me, Kate* and *South Pacific*, had done more to bring listeners and record buyers to the 33⅓ than any amount of chin-strokey classical music or heavy opera.

This wave of popularity made Lieberson the leading, go-to producer for cast albums, and cast albums and movie soundtracks became the leading, go-to money spinners of the day. Musicals weren't all Columbia's territory, by any means. The eventual movie soundtrack of *South Pacific* would be a massive hit for RCA Victor, for example, and the soundtrack to *The Sound of Music* would become one of the most successful soundtrack albums in history. Decca too cemented a reputation for its full cast recordings. But Lieberson continued to land smash after smash, including the original cast albums for *Gentlemen Prefer Blondes*, *Cinderella*, *West Side Story*, the first complete recording of *Porgy and Bess*, *On the Town* and *My Fair Lady* – the original Broadway cast in mono in 1956, the original London cast in stereo in 1959. His last Broadway cast recording came in 1975 with *A Chorus Line*. Writing in the liner notes of the double LP *This is Broadway's Best* in 1961, he described how a musical denuded of all that made the live performance what it was – plush sets, bright colours, costumes, dancing – only helped it 'shine forth with a brilliance never before suspected'.

Books, like records, used to be sold in plain buff wrappers, designed to protect the book on its journey between printers and booksellers. These wrappers started carrying printed text – normally a summary of what else the publishers had to offer. The earliest proper dust jackets, that is jackets that not only protected the book but also carried some kind of unique design specific to that book, began to appear around the turn of the century. Two famously rare and fantastically valuable examples were the jackets for the first UK editions of *The Wind in the Willows* and *The Hound of the Baskervilles*. Both were buff, with simple one-colour designs that mirrored what was on the book's front cover board – a figure in some reeds for *Willows*, a silhouette of a dog for *Baskervilles*. They're rare because they were made of fragile paper, and they came at a turning point – when the average book buyer would tear off and discard the protective cover. As a result, there are only two or three surviving examples of each, which are worth hundreds of thousands of pounds today.

Something similar occurred with records. Some might gently mock Edison – for the fact that he not only continued to make the cylinder after everyone else had left the business, but also that he insisted on packaging his beloved Blue Amberols in tubes that bore his name, his face and his patent information, with nothing – or, at least, next to nothing – about the artist or performer. In fact, however, he was only doing what they all did. For generations discs came in plain sleeves, sometimes completely blank, others with simple Art Deco designs, others covered in branding for that particular label or type of record. Some of these designs are beautiful. But considering that by the middle of the century the book's once humble dust jacket had become such a vital part of its identity, it seems strange that it took so long for records to follow suit. Especially when you consider the contemporary sheet-music trade – these would frequently have all sorts of images and designs, often accompanied by photographs of the conductor, famous performer or composer. Yet still, records stubbornly refused to come dressed in anything other than their familiar, muted, utilitarian clothes.

Longer pieces and proto-albums would be recorded across multiple 78s. By the 1920s, more record labels had begun offering their special record albums in dark-coloured books, with leatherette or canvas bindings. Using these, collectors could group together batches of records by an artist or type – foxtrots in one, waltzes in the other. One of the first proto-albums came from German record company Odeon, which issued the *Nutcracker Suite* on four double-sided discs in 1909. By the 1930s, however, more companies were expanding on the album idea, issuing especially preassembled albums – multiple discs of recordings from an artist, genre or suite. However, these collections generally all looked the same, and only very occasionally did some special releases come with any kind of design fanfare, perhaps a simple piece of artwork pasted on the front.

This workaday approach was challenged in the 1940s through the pioneering work of Alex Steinweiss. Steinweiss was born in New York in 1917, the son of Polish immigrants. After studying graphic design at school, he earned a scholarship to Parsons School of Design in New York and graduated in 1937. He worked as an assistant to Joseph Binder, the Viennese poster artist, which would have an enormous influence on his approach. And then he landed a job with Columbia, becoming their first art director-type person, working from an enormous room he called 'the ballroom', where he would produce booklets, posters, catalogues and anything else vaguely promotional. And to him, the drab, largely buff record sleeves seemed such a wasted opportunity.

He approached his managers with the obvious-with-hindsight idea of adding some colour to the sleeves. He assumed this would be met with an immediate 'no' because of the increased costs, but it wasn't. He began work on what is generally thought of as the first album cover in 1940. It was for a Rodgers & Hart collection, a simple, theatre marquee-type design, with the title in lights, over a background of record grooves in red.

Over the next few years he designed every single album cover Columbia put out. Keep in mind, these were still the 78-era multi-disc albums, so big, hefty objects. He used flat, bold blocks of colour. Without any type suppliers nearby, he tended to hand-letter all the titles himself, developing what would become known as the Steinweiss Scrawl. If you find some of his work now, it may put you in mind of wartime propaganda, of celebrated Penguin books, of the film titles of Saul Bass.

He was working alone, and breaking new ground, and so was able to try out all kinds of new things. His cover for the original Broadway recording of *South Pacific* (1949) has been in almost continuous use ever since. A notable early example was the Columbia Masterworks M 534 – *Paul Robeson's Songs Of Free Men*, issued in an album of four shellac 10-inch discs in 1943. This shows an arm, with a broken chain still manacled to the wrist, stabbing a knife into a snake bearing the Nazi Swastika.

After the war, by now working freelance, Steinweiss was invited to a meeting with label head Edward Wallerstein, who seemed to be acting oddly. With a furtive air, he put a record on a turntable. Steinweiss listened, waiting for it to end, waiting for the inevitable turnover, but it didn't happen. As you might have guessed, he was being shown the first LP, and Wallerstein told him that these newfangled microgrooves were giving him a new headache – packaging. The grooves were too fine and delicate for the tough old sleeves and envelopes used for shellac 78s. He asked Steinweiss to tackle the problem.

The very first 12-inch LP, a Mendelssohn violin concerto played by Nathan Milstein, was issued with a Grecian column design, printed white on dark blue, borrowed from an earlier 78rpm album cover. Steinweiss would eventually develop thin boards covered with printed paper, which could be easily folded and glued. The industry-standard LP cover became four-colour printing on the front, black-and-white on the reverse. The delicate microgroove required further care, so they would eventually add an inside record sleeve. Again, these were generally paper, very thin cardboard or rice paper, sometimes with an inner liner, sometimes with a centre hole so you could see the label. (These too would eventually become a canvas for more artwork, lyrics, sleeve notes and photos.)

The fashion soon began to shift from graphics and imagery towards portraits of artists and performers. That's not to say Steinweiss's work dried up (by the 1950s he had begun long associations with Decca, RCA, London Records and Remington), just that the look of pop records was changing.

This book is called *Into the Groove*, not *Between the Covers*. So while we could spend more happy hours examining the work that came after Steinweiss, exploring the impact of Jann Haworth and Peter Blake, Andy Warhol, Storm Thorgerson, English art design group Hipgnosis, Pennie Smith and Ray Lowry, Peter Saville, Jean-Paul Goude or Jamie Reid, I'm not going to do that. There are a multitude of illustrators, artists, designers and photographers who picked up the baton, creating pop culture iconography as familiar today as a dollar bill or the Eiffel Tower. As the 1950s gave way to the 1960s, the photographs of angelic popsters, clean-cut harmonisers, smiling Beatles and pouting Stones were giving way to artistic expressions, pop-art statements and gimmicky innovations, hypnotic gatefold vistas and shaped cut-outs, illustrated inner sleeves and full-colour lyric sheets, punk pastiches and multi-disc glam excesses. Steinweiss was a conduit, a kind of Lonnie Donegan of record design, whose brush, pencil, scalpel and style helped change the way companies packaged and we consumed popular music, turning the humble protective sleeve into a canvas, a call to arms, a tribal banner for fans to gather beneath. And whatever your opinions on the competing advantages of lengthy cassettes, hardy compact discs, minidiscs, laser discs or lossless streaming, these formats can't rival the pleasures of a brand-new, full-sized LP by your favourite band, unwrapped from its cellophane for the first time, ready for several minutes of detailed examination.

TWENTY-ONE

Stereo and Magnetic Tape

'Home taping is killing music ... and it's illegal.'

British Phonographic Industry anti-infringement campaign, October 1981

Magnetic recording has been with us almost since the start of this story. It was in Edison's original patent (he imagined a kind of steel sheet), was extensively tested at the Volta Lab, and had been played around with on and off for years before it was rendered workable and scalable by some ingenious Germans.

The magnetic era of recording runs from 1945 to 1975, and while tape would mount a massive threat to vinyl's dominance, it also helped the vinyl record become even better than it was already. The father of magnetic recording was a Dane named Valdemar Poulsen. He invented a device called the telegraphone, which had a steel wire wound round a drum. An electromagnet travelled along the wire to record the signals, which it could then replay as it made a return pass. At the 1900 World's Fair in Paris, Poulsen recorded the voice of Emperor Franz Josef of Austria, believed to be the oldest surviving magnetic audio recording. Then, during the First World War, telegraphones were used to record German-sent telegraph messages.

Magnetic recording really took off in 1935, when BASF and AEG in Germany both developed a means for coating paper (then film) with magnetic material. This led to the first generation of Magnetophon recorders, which were being used by German radio stations.

American audio engineer John T. Mullin served in the US Army Signal Corps during the war and was assigned to find out everything he could about German radio and electronics. A visit to a studio at Bad Nauheim near Frankfurt resulted in him getting hold of a cache of AEG Magnetophon recorders and tapes, which he had shipped home. After the war, he set to work improving them with the aim of securing interest from Hollywood studios.

Public demonstrations were a huge success. Soon the Ampex Electrical and Manufacturing Company had created their own version – the Model 200 tape recorder. This would become the standard, go-to in-studio tool for major radio stations and networks. It was so easy to use, and it sounded so good. Bing Crosby was in constant demand as a live radio performer but was growing tired of having to turn up week in, week out to perform. The radio bosses weren't keen on pre-recording their most popular star on transcription disc, as the audio quality wasn't good enough. The new tape format set Bing free. His season première on 1 October 1947 was the first magnetic tape broadcast in America.

Pre-records, re-records and edits were relatively quick and easy, and broadcasts were suddenly a lot slicker. Even bigger changes were coming to the recording groove. Magnetic tape allowed audio engineers to record multiple segments of audio and keep them perfectly in sync with each other. Multi-track audio also allowed for the advent of stereo sound, as engineers started recording with a 'stereo image' in mind. Tape, and the quality of tape, meant that it was suddenly much easier to record high-quality masters to commit to disc or to sell to other labels.

Sam Phillips had used an Ampex 350 tape machine to record 'Rocket 88', which he then sold to Leonard and Phil Chess in Chicago, who released it as the 78rpm Chess #1458. The sale of this master tape helped Phillips fund Sun Records. Elvis's recording of 'That's All Right', released in July 1954, was recorded on two Ampex 350 recorders, which Phillips then used to create the trademark 'slapback' audio delay. Guitar legend Les Paul was one of the first to push the boundaries of multi-tracking, innovating tape delay, phasing and overdubbing. His first and most influential step was the 'Sound on Sound' recording, which he created by added a second recording head to an Ampex tape recorder, allowing him to play along with a previously recorded track and then mix the two together on a new track.

Two eyes create a perception of depth, and stereo can give a sense of presence. Recording music with two appropriately positioned microphones, then playing that back via two separate loudspeakers, may sound like an obvious idea, a concept we all now take for granted. But the promise of stereophonic recording and reproduction was a dominant theme in the improvement of the long-playing record. First, how? For years the stylus, pick-up and cartridge only had to handle one channel. How can a single groove suddenly handle two channels? Where do you put the other one?

Stereo also traces its history right back to the start of this story, to the carbon microphone, which was independently developed by an Englishman named David Hughes, by Emile Berliner (as we already know), and by Thomas Edison (who you also might remember, and who was awarded the patent in 1877).

In 1881, a telephonic transmission system was demonstrated by a French inventor named Clément Ader at the International Exposition of Electricity in Paris. He had arranged 80 telephone transmitters across the front of a stage at the Paris Opera to create a form of binaural stereophonic sound. This he connected, via the first two-channel audio system, with a suite of rooms at the Exposition of Electricity. There, visitors could hear performances in stereo using headphones. During these experiments, it was demonstrated that the stereo effect meant it was possible to follow the movement of singers on stage. And the experiment was further commercialised some years later with Théâtrophone, a telephonic distribution system available around Europe, which allowed subscribers to listen to opera and theatre performances over telephones.

More recently, archivists and record collectors have unearthed cases of 'accidental stereo'. In 2014, a remarkable compilation by Mark Obert-Thorn and Andrew Rose brought together all known 'accidental stereo' classical recordings from the 1930s – where the original 78 is brought together with standby recordings that were made at the same time using another mic.

Elsewhere, Harvey Fletcher at Bell Laboratories was investigating 'wall of sound' techniques, using lines of microphones hung in front of an orchestra. Each mic fed to a corresponding loudspeaker in an identical position in a separate listening room. He also conducted stereophonic test recordings, using two microphones connected to two styli cutting two separate grooves on the same wax disc. This experimental recording was made with Leopold Stokowski and the Philadelphia Orchestra at Philadelphia's Academy of Music in March 1932 and is the earliest known (intentional) stereo recording to survive.

Meanwhile, British EMI engineer Alan Blumlein designed a way to make the sound of an actor in a film follow his movement across the screen. In December 1931, he submitted a patent application and over the next few years developed stereo microphones and an ingenious stereo disc-cutting head, recording a number of short experimental films with stereo soundtracks. The Blumlein patent covered many founding ideas in stereo: a pair of velocity microphones with their axes at right angles to each other, which is still known as a 'Blumlein Pair'; recording two channels in a single groove, using the two groove walls at right angles to each other and 45 degrees to the vertical; and a stereo disc-cutting head. While Blumlein had originally been inspired by, and so was most focused on, the use of stereo in film, he also recorded Mozart's 'Jupiter' Symphony, conducted by Sir Thomas Beecham at Abbey Road Studios, using this vertical-lateral technique, in 1934.

In 1937 Bell Laboratories demonstrated a form of two-channel stereo sound that worked by having two optical tracks on film. Soon movie studios followed suit, developing three- and even four-track systems. The first big-ticket stereo film was the MGM Judy Garland vehicle *Listen, Darling* in 1938. Then Bell/Walt Disney upped the ante with *Fantasia*, using the new Fantasound sound system, which employed a separate film for the sound, with four double-width optical soundtracks – three for left, centre and right audio, and a fourth 'control' track. You can imagine that this was state of the art and so required state-of-the-art machines. As a result, Fantasound was only ever exhibited with the movie as part of a US-only roadshow – regular movie-goers saw it in standard mono.

Vinyl as a format for records was in use for some time before it was made more widely available to the public. Similarly, although multi-track and stereo recordings existed, it took time for them to percolate down to the domestic record buyer. Stereo was a specialist market, with EMI, HMV and Columbia all putting out classical stereo tapes for playing on domestic machines, but the problem of how to cut and read two channels in vinyl had not yet been resolved.

One of the first to try it on record was Emory Cook. He had previously released a number of stereo demonstration recordings on reel-to-reel tape. He tried cutting unique stereo discs, where there were two completely separate sections of groove for each channel, one cut after the other. These were played back by a special two-headed pick-up, without an inch gap between the two styli. Then, in 1957, Westrex, the Western Electric offshoot, created a new type of single-groove 45/45 cutting lathe for LPs, which required a single, newly developed stereo cartridge to trace the two channels, which were cut at 45 degrees to one another in a single groove. The new technology was paraded before the major studios, who were also given new Westrex cutters to experiment with.

While they were still pondering what to do, a man named Sidney Frey got there first. Frey ran a relatively obscure New York label called Audio Fidelity Records, which had been around since 1954 and generally put out diverting sonic spectaculars. In October 1957, Westrex cut masters of some of Audio Fidelity's stereo recordings. Side one had the Dukes of Dixieland, a New Orleans Dixieland-jazz revival group, while side two was railroad sound effects.

The masters were really little more than test cuts – they did show what the new technology could do, but they had imperfections and noise issues, which Sidney, seeing an opportunity for publicity and in a great rush to grab it, decided not to worry about. This was a plain demonstration disc, so it came out in a plain black sleeve with a simple gold label explaining that this was a 'stereophonic demonstration record' for 'test and laboratory purposes'. Five hundred copies were pressed, and it was introduced to the public on 13 December 1957 at the Times Auditorium in New York City. Frey then took out an advertisement in *Billboard*, promising a free copy to anyone in the industry who wrote to him on company letterhead.

This proved to be the making of Audio Fidelity, which survived as a specialist purveyor of audio spectaculars in the new glossy era of stereo. It was the leader in a rapidly growing niche market. By June 1958, the price of the special stereo magnetic cartridge required for playing the discs came down from $250 to the much more reasonable $29.95. The first four stereo discs available in the US all came courtesy of Audio Fidelity: *Johnny Puleo and His Harmonica Gang*'s self-titled album and *Marching Along with the Dukes of Dixieland Volume 3* in 1957, and *Railroad Sounds: Steam and Diesel – the Sounds of a Vanishing Era* and *Lionel* by Lionel Hampton and His Orchestra in 1958. Then, in the summer of 1958, Audio Fidelity recorded another batch of classical LPs in Walthamstow Town Hall, London. The orchestra was the specially formed 'Virtuoso Symphony of London', which consisted of London orchestral players and leading instrumentalists including Anthony Pini, Frederick Riddle, Reginald Kell and Marie Goossens.

So there you have it, the first true stereo was revival jazz and train sounds, imperfectly cut. Again, there was a long gap before stereo had become mainstream. Until the mid-1960s, record companies mixed and released most popular music in monophonic sound. From the mid-1960s until the early 1970s, major recordings were commonly released in both mono and stereo. And ever since, recordings that were originally released only in mono have been released in 'pseudostereo' using various techniques. Mercury Records engineer C. Robert Fine had pioneered a monaural recording technique in 1951 that was branded 'Mercury Living Presence'.[27] Here they switched to a three-channel stereo recording, using a main central microphone, with two side mics adding depth and space. RCA had 'Living Stereo', Decca had Phase 4 Stereo and its progressive Deram Label, where records were made using the panoramic 'Deramic Sound', designed to show off a more dynamic stereo field.

Meanwhile, tape was mounting a serious threat. Blank and pre-recorded tape began to be sold to the public in greater numbers from the mid-1950s. This started with some pricey 7-inch reels that played at 19cm per second, but it really began to attract audiophiles when EMI's Stereosonic tapes went on sale in 1954.

To make tape cheaper, companies began reducing playback speed by half, then switching from half track to quarter track – half track used the entire width of the tape to play or record, quarter track used half the width of the tape, so would play or record in both directions.

The old reel-to-reel tape remained a fiddly little blighter – you had to thread the loose end on to a second spool, which was fine once you had the knack but certainly wasn't a quick operation. To overcome the whole business of spooling, manufacturers tried various means of repackaging the tape – in different designs of magazine, cartridge or cassette. The 8-track tape (formally called Stereo 8) was popular from the mid-1960s. Then, at the 1963 Berlin Radio Show, Philips

---

[27] These recordings were remastered in the 1990s by the original producer, Wilma Cozart Fine.

turned up with a portable battery-powered recorder. This device used a 'compact cassette' that could fit in the palm of your hand. Inside was a tiny spool of tape, designed to run at about 5cm per second. Not only that, it sounded pretty good. Soon many other companies were making their own machines to play compact cassettes, and original 'Musicassettes' were coming on sale. By 1977, the year of vinyl's centenary, cassettes were taking an enormous bite out of the record's 100-year head start, and by the 1980s, 90-minute cassettes had all but killed off the old 8-track cartridge. By the time Cliff Richard appeared in his promotional video for 'Wired for Sound' in 1981, roller-skating about with a Walkman on his hip, the LP record seemed doomed.

TWENTY-TWO

Discos and Curios

'The combination of all three in unison gives an intensity of volume sweetness and richness of tone which seem almost beyond belief.'

Multiplex Graphophone Grand advertisement, 1901

A Magnetic cartridge, or pick-up, is an electromechanical transducer that is used to play records on a turntable. It holds a removable or fixed stylus. As the stylus tracks the serrated groove, it vibrates a cantilever. You can have a 'moving magnet' or 'moving coil' cartridge. With the first, the cantilever holds a magnet that moves between the magnetic fields of sets of electromagnetic coils in the cartridge. With the second, the coils are on the shifting cantilever, the magnet fixed in the cartridge. With both, the shifting magnetic fields generate an electrical current in the coils, and the signal can be amplified and converted into sound. The 'London Decca' cartridges had fixed magnets and coils. The stylus shaft was composed of the diamond tip, a short piece of soft iron and an L-shaped cantilever made of non-magnetic steel. Since the iron was placed very close to the tip, the motions of the tip could be tracked very accurately. Decca engineers called this 'positive scanning'.

Tracking force is the weight at which the stylus sits on the record. This needs to be at the correct weight to get the best sound and prevent damage to the grooves. If it's too heavy, the stylus will push down and cause distortion and damage. Too light, it will skip across the record, scratching the vinyl. The tonearm itself supports the cartridge in the correct position over the record, holding it at the proper angle to correctly read the stereo channels in the groove.

Some arms are built for one machine, others are designed to work on any turntable, adjustable to work with different heights of platter and different cartridges. Whether old Bakelite, metal or carbon fibre, the tonearm has to apply the correct amount of weight to the stylus to stop any skating towards the inside of the record, and of course it carries the cabling that transmits the signal from the cartridge to the amp. The tonearm, then, has to be relatively light but also stiff and strong, and yet free to move – held too stiffly, and this will put stress on the stylus and groove, impacting the signal. At one end sits the cartridge, at the back end you'll find adjustable counterweights. As tonearms developed, more and more complex and adjustable weights and counterweights were introduced – although Blu-Tack and coins remain popular.

This is the area where vinyl enthusiasts really can outgeek each other. The reason my parents were so tense around us children when we were anywhere near the turntable was that a sticky, jammy hand in the wrong place really can mess up the whole thing. There's the gimbal, for example – the pivoted support that allows the arm to rotate. Most tonearms had sets of gimbal bearings, which had to be polished to a high surface-finish in order to offer the minimum friction. Any play in the set-up could allow the tonearm to chatter – to rattle. The 'unipivot' design, the first put into production and made by Cosmocord in the 1930s, was introduced to reduce this still further by replacing all the gimbals with a single bearing on which the entire tonearm, counterweight and cartridge rested. The most popular unipivot design came from Frank Ockenden and John Wright. The pillar was mounted on the deck, with a pivot. This went into the lower face of the hub – sitting within a bath of viscous silicon fluid, which gave the whole delicate set-up enough torsional stability. Without the bath, the tonearm could bounce about all over the place. This kept it in check, but not too firmly.

This, then, is the difficult balance that is continually in play and continually being tweaked and adjusted to this day. The German firm Clearaudio produced magnetic bearings that 'float' the arm over the record with no point of contact at all. Meanwhile, to the layman, the promise of the laser turntable, one that uses a beam of light to read the groove, seemed to be the ultimate solution. Surely these would be able to perform the function of the stylus without any weird resonance or chatter, without any tracking errors or wear, and could read the tiniest undulations in the engraved waveform? Surely distortion and noise would be dealt with once and for all, and the dynamic range and frequency response would be wider than ever? Surely they could offer sound without compromise?

The origins of the laser turntable, like so many things to do with vinyl, can be traced back to May 1977. William K. Heine was presenting a paper on 'A Laser Scanning Phonograph Record Player' to the 57th Audio Engineering Society convention in Los Angeles. In his paper he detailed what he had been working on since 1972 – a single helium–neon laser that could track a record groove and reproduce the audio. With his working prototype, the 'Laserphone', Heine had high hopes that soon lasers for phonographic playback would become the norm. A few years later another company released the Finial LT-1. This worked but never went into full-scale production, being far too expensive, and was eventually sunk by the arrival of the Compact Disc. The Finial investors eventually sold the patents to a Japanese maker, who in 1997 made the ELP LT-1XA Laser Turntable, a snip at $20,500.

Considering laser turntables had been around for so long, why have they not become the norm? The simple answer is that they work *too* well. The clumsy diamond stylus is like a spaniel in the groove in comparison to laser. And just as a spaniel will clear the kitchen floor of crumbs, so the hard physical stylus will clear lots of the worst debris out of the way as it goes. The laser can't do this, and so picks it up and amplifies it instead. The result is the record about to be played by laser has to be fully wet-cleaned. A 2003 review in *Stereophile* noted clear vinyl, which would have been virtually silent with a normal stylus, would sound like someone munching crisps when played with a laser.

In fact, the 1970s was something of a golden era for this kind of thing. The decade saw a succession of world-changing audio level-ups, aimed at the well-heeled audiophile, which ultimately failed to catch on but left an impression.

In the wake of stereo came the first quadraphonic records. This technology was achieved by electronic matrixing, where the additional channels were combined into the main signal and sent packing into the walls of the stereo groove. When quad records were played, phase-detection circuits in the amplifiers would decode the signals into four separate channels. Then there was noise reduction[28] – dbx-encoded discs, which were incompatible with standard pre-amps, were recorded with the dynamic range compressed, and then dbx automatic gain control reduced any surface noise even further. This was joined by the short-lived Telefunken/Nakamichi High-Com II noise reduction system in 1979, then the lower-end CX 20 noise reduction from CBS in

---

[28] Back in 1963, RCA Victor launched Dynagroove. This too was a new recording system designed to reduce surface noise, but it was not popular with audiophiles. It's notable for being one of the first times computers were used to modify an audio signal. It used a new material with less surface noise, and to manually boost bass and reduce high frequencies within certain parts of the music, which, it was hoped, would translate well through the cartridges of the day.

1981. Another was 'direct-to-disc', which bypassed magnetic tape in favour of a 'return-to-roots' approach of recording to master lacquer discs.

At the other end of the scale were various 16rpm records. These span too slowly for proper high fidelity. However, this speed increased capacity, making them perfect for spoken-word records, pre-recorded broadcasts and the legendary Seeburg 1000 record player, which was designed to play 9-inch 16rpm discs for background Muzak in public spaces, where sound quality wasn't paramount. And for every new innovation, there were rebrands disguising backward steps. The push for ever-higher fidelity rubbed up against the need to cut costs. Sound was dependent on the quality of vinyl, and from the 1970s, discs became thinner. The difference between an LP from the late 1960s and a cheap 1980s reissue of the same album was enormous. Some seemed to arrive at the shops ready-warped.

The character and meaning of the EP, or extended play, changed over the years. These originally sat somewhere between the LP and the 7-inch single – essentially lower-quality 45rpm discs that could therefore squeeze on four or five tracks rather than just the one. RCA Victor introduced their 'Extended Play' 45s during 1952, the narrower grooves made by lowering cutting levels and sound compression. Some labels began releasing albums as collections of 45rpm EPs, but from the 1950s more and more contained compilations of singles or album samplers. The term 'EP' eventually came to mean any mid-length release, whatever the format.

As part of the disco explosion, Motown Records introduced its own 'Disco Eye-Cued' discs. These records were designed specifically for the club, where particular segments of the track – drum-breaks or choruses perhaps – were indicated by more widely separated grooves. This gave the DJ a visual cue about where to place the needle, and soon the New York label AVI Records followed suit, introducing expanded grooves on its disco discs, the first being Captain Sky's 'Wonder Worm'.

For generations, the humble flexi-disc would be the entry point to record collecting. These wafer-thin, flexible sheets with moulded grooves were often attached to magazines or fanzines as giveaways. From the early 1960s Japanese monthly *Asahi Sonorama* began including inserted 'Sonosheets'. The most famous and avidly collected examples were the special Christmas recordings The Beatles made to their fans from 1963, which were filled with messages, music and comic pastiches. At the other end of a scale marked 'musical merit', the 1980s UK comic *Oink* came with a hellish flexi of singing pigs. During the 1980s home-computing boom, flexi discs full of data even graced computer magazines. The discs of video games could be fed into home computers via ordinary turntables – either directly or you'd have to copy the weird rasping sound to a cassette and use that. One 1984 issue of *Computer & Video Games* had a Thompson Twins-themed *Adventure Game* encoded into the grooves. Around the same time musician and comedian Chris Sievey, better known as his comic persona Frank Sidebottom, issued a 7-inch single called 'Camouflage', with his own home-coded game, *Flying Train*, on the flipside.

Other throwaway curios include the Stollwerck chocolate records that came with a tiny tin gramophone in the 1900s, and the even tinier playable stamps that were created by an American entrepreneur named Burt Todd in the early 1970s at the request of the Bhutanese royal family. The talking stamps could be stuck on a letter or played on a turntable, or both, and included the Bhutanese national anthem, folk songs and a (very) concise history of Bhutan.

Record engineers and cutters were in the habit of signing off creations with matrix numbers issued by the record company in the run-out groove. These might include letters representing the year, genre or imprint, the master tape size and speed, a description of the release (whenever mono or stereo), and the studio code. Some engineers were in the habit of adding signatures, initials, doodles or longer messages. John Lennon's engineers were encouraged to etch into the blank spaces, while George Peckham used to sign off masters with his nickname 'Porky' or perhaps 'another Porky Prime Cut'. When it came to the final cutting of the Voyager Golden Record at CBS, producer Timothy Ferris and engineer Vlado Meller decided, relatively last-minute, to add the inscription: 'To the Makers of Music – All Worlds, All Times.' A beautiful sentiment for a space-faring record, and one that almost stopped the space-faring record going to space. You see, the golden discs were made, the project was complete and Carl Sagan had his feet up. The discs were all ready to be fixed to the sides of the space probes, ahead of launch in August and September 1977. A NASA box-ticker was checking them against a form. On the form was a detailed description of the records' physical characteristics. The form didn't mention an inscription, so it was declared a non-standard part and discarded. They even started to replace it with a black disc to make up the weight. Luckily, Sagan got wind of the trouble and managed to persuade them to let it pass.

There's always that bit when the very end of the spiral groove meets its own tail and keeps the stylus spinning in place. This locked groove stops the stylus going onto the paper label, and every now and then studios have played around with these too – most famously a slightly creepy bit of endlessly repeating audio at the end of *Sgt. Pepper's Lonely Hearts Club Band*.

A popular creation myth from the world of hip-hop is how Grand Wizzard Theodore invented scratching. He was 12 at the time, and already a popular DJ. He was at home in the Bronx when his mum burst in to tell him to turn the music down. He stopped the record with his hand, and the sound inspired him to try scratching in public for the first time. The date was 18 August 1977, and he debuted the technique during a set at the Sparkle club, using 'Bongo Rock' by the Incredible Bongo Band. The technique took off with direct-drive turntables. Belt-drive turntables were unsuitable for scratching. For the technique to work, you needed the disc to come back up to playing speed immediately. If you stopped a record with your hands on a belt drive, there would be a slow return to the playing speed. Theodore experimented with and perfected scratching using a Technics SL-1200, a direct-drive turntable released by Matsushita in 1972. A century after Edison, and the phonograph had become its own musical instrument.

TWENTY-THREE

Revolutions in Space

> 'Mary had a little lamb, its fleece was white as snow, and everywhere that Mary went, the lamb was sure to go!'
>
> Conrad Beneshan, 1977

In August 1977, a centenary celebration took place at Thomas Edison's former home in West Orange. The highlight of the morning's activities was a recreation of the first successful phonograph playback by a long-serving company technician named Conrad Beneshan. Before about a hundred guests, he bellowed 'Mary Had a Little Lamb' into a near-perfect replica of the original phonograph. Beneshan had done this before, plenty of times, but he was still a little nervous doing it in front of a crowd on such a momentous day. As we know, the machines were notoriously unreliable and, as he told the crowd, the foil was just plain old 'supermarket stuff'. He began cranking again, and when his own voice came back out through the megaphone, he exclaimed with relief, smiling at the cheers and applause that erupted from the audience. This event took place on 12 August 1977, when the launch of the Voyager 2 space probe was about two weeks away.

For me, the Voyager Golden Record on the heels of this tin-foil recreation is like a fighter jet flypast at a centenary of the first heavier-than-air machine. That's not to say that we should think of the Golden Record as the greatest ever made. It's not. For a start, it was designed to be played at 16rpm, and at that speed audio fidelity is compromised. However, the story of how it came together, and the fact that it will outlast our planet – and probably our solar system – means that it is not only the fastest, the furthest and most hard-wearing record, but also the most perfect assimilation of hope and ingenuity, the format and the grooves pushed to its limit, and all done on the minuscule off-chance that it will ever be discovered by extraterrestrial intelligent beings.

The idea for a record may seem odd. Why put a record in space? The answer to that question has three strands. First off, this was the second message to be fixed to the side of a NASA space probe. A few years before Carl Sagan, his second wife Linda Salzman Sagan and astronomer Frank Drake had worked on a plaque that was fixed to the side of the Pioneer space probes. These were the first objects to visit Jupiter and Saturn and, because of gravitational slingshot, they were destined to be the first human-made objects to leave our solar system and head off into deep space. With the magnitude of this idea came the momentous thought: surely we need to send something with them that explains where they came from – some kind of 'made on Earth' stamp. Or something.

Carl Sagan knew an inspiring idea when he heard it. So, together the trio designed a postcard-sized plaque that was engraved with nude figures of a male and a female drawn by Linda, a kind of ingenious star map by Frank and a planetary fingerprint of our solar system by Carl. It basically said: 'We were the ones who made this spaceship, and this is where we live'. The plaque was a huge success, generating lots of interest in the project.

When the Voyager mission came round a few years later, it seemed a good time to try something like this again. This was the more advanced mission, dubbed 'The Grand Tour' – a full tour of the outer planets, with Voyager 1 off to Jupiter and Saturn, Voyager 2 visiting Jupiter, Saturn, Uranus and Neptune. Towards the end of 1976, NASA approached Sagan and asked: would you like to do another plaque-type thing?

Carl and Frank put their heads together. For the second message, they didn't want to do the same thing again, they wanted to do it better. Frank had been at the heart of the search for, and communication with, extraterrestrial intelligence for a generation. His view was that with any interstellar messaging, you needed to treat aliens the same way you'd treat babies or people who don't speak your language – you use pictures. Meanwhile, Carl was asking his circle of friends what they thought they should do this time round. One of the suggestions that came back was they should send a reel of tape aboard the Voyagers, containing a snatch of Beethoven. This got Carl thinking about music.

Frank, meanwhile, a problem-solving engineer at heart, was looking at how to cram more information onto a modestly sized piece of metal. How can you force a metal plaque to contain more data? You must increase the surface area, he thought. And how can you increase the surface area? *Grooves.* Records have a hugely increased surface area because of all those lovely microgrooves. Then he thought of *pictures* and thought … you could put pictures on the record. You could convert video signals into sound and encode them into the grooves.

Wait. Hold up. What?

Yup, the Golden Record has not only music and sounds but also images – about 120 of them, in colour and black and white. These were carved into the grooves as analogue video signals converted into sound. The actual sound, which you can listen to now,[29] is a rather nasty rasping beeping. Which leads us to the amusing thought of an alien technician, lightyears away and many millions of years in the future, listening to the Golden Record, enjoying a bit of Bach and Beethoven, then wincing as the image audio cuts in, muttering: 'This is a little too avant-garde for me …'

Anyway, back to 1977.

---

[29] To listen, go to www.youtube.com/watch?v=ibByF9XPAPg, or alternatively search for 'First Ever Decode of Voyager Audio Images, in Real Time'.

There's a piece of paper from around the time Frank had his idea for an interstellar record, on which he jotted down a very early proposed playlist. It contains around 12 items, including sound effects, some atmospheric music and a picture of the Taj Mahal. This illustrates that for Frank, a high priority for the record project was that it include images. Carl, meanwhile, was absolutely beside himself with the idea of a record that included music, the idea of sharing human art with the universe, and all the interesting questions and challenges it posed – what do you choose? What music represents an entire planet?

In the end, he and his team of helpers went all out. Fairly early on, they decided to go for 16rpm, as it just was too painful to make the playlist any shorter. With 90 minutes at their disposal, they had enough elbow room to show off what we humans can do. Another late time-saving decision was to split the weird image audio over left and right channel, to save run-time. They amassed greetings in multiple languages, whale song, a sound essay describing the evolution of life on Earth, a strange audio-fingerprint of the orbits of our solar system, a full EEG reading of someone's brainwaves, then images of birds, animals, machines, diagrams and, of course, more than 90 minutes of music. The music includes classical smashes, opera, intimate field recordings, world music, a breathtaking raga from India, an epic, atmospheric guqin piece from China, and popular Western fare headed by Chuck Berry, Louis Armstrong and Blind Willie Johnson.

Just a century on from Edison managing to fit a few seconds of a bellowed nursery rhyme onto some tin foil, Sagan was imprinting symphonies in gold.

The cutting was done in New York by audio mastering engineer Vlado Meller, who had been with CBS since 1969. Now, imagine you work at CBS in the mid-1970s. You're at the top of your game, you're cutting some of the best-sounding records ever put out. Bruce Springsteen and Meatloaf are just down the hallway. Then, some geeky space cadets turn up with 90 minutes of music, tapes of sound effects and strange audio pictures, and they want you to fit it all onto one disc. Perhaps this explains why the record's producer, Timothy Ferris, on first approaching one of the CBS higher-ups was greeted with the words: 'So let me see if I get this straight. You made this piece of shit and it's going to go whirling around the Earth, and for some reason I've gotta clear this motherfucker?'

Meller, meanwhile, just thought it was weird – why make a record and send it into space? Who's going to hear it?

Cutting 29-minute sides was tricky enough, but 40 to 45 minutes was a very delicate operation. The grooves are so packed, so thin, that conventional turntables can't track them, because when it starts spinning the arm just slips off – it's pulled to the side. With such a super-thin groove, you have a chance of skipping. So they had to do lots of test cuts. Luckily, as there was a fair amount of spoken word, sound effects and animal noises, it meant they could save space, opening up the grooves for the music.

While Sagan is the famous figurehead of the project, there were many others who got it over the line: Ferris produced and brought it all together, Carl's wife-to-be Ann Druyan amassed sound effects, found records and audio tapes, helped compile the playlist and made sure Chuck Berry was aboard, artist Jon Lomberg found images, suggested music and created unique artwork for the picture sequence, Linda Salzman Sagan helped bag all the greetings and languages, and Frank Drake, aside from coming up with the idea in the first place, wrote and designed the how-you-play-a-record instructions for aliens, essentially using line drawings, binary notations and a diagram of hydrogen atoms to describe which direction the record was supposed to spin, how fast and how to rebuild the video images from the audio.

Timothy Ferris left the CBS building with the freshly cut lacquers under his arm in June 1977. He took a cab directly to the airport, caught the red-eye to Los Angeles and hand-delivered the discs to the James G. Lee Record Processing Center in Gardena, California. Here, the sides became copper mothers, which were then bonded together to form single discs. These were gold-plated, encased in aluminium and mounted on the sides of the spacecraft, ready for their long journeys. Really long. They've been going for more than 40 years, and they've only just left the solar system.

There is something odd about how vinyl keeps getting back up. It's written off, written off, written off again. It was struck down by radio, only to pick itself up and steal some of radio's powers to make itself stronger. Then it did the same with film, then tape. Every time it was knocked down it learnt something and got better at being what it was – a receptacle. The nearest it came to actual death was probably in the 1990s, with CDs firmly in the saddle, cassingles still being made, mini-discs in town and MP3 players approaching. Records had been relegated to a fringe item, and most bands and artists had stopped bothering with vinyl at all.

Tapes were excellent. In the 1980s, nothing else could reproduce so much music with so much ease. But they had several drawbacks. They got tangled up in your car stereo, and the endless forwarding or rewinding to find a song was annoying. Still, we all loved tapes, mainly because of the heady power of being able to make our own mixtapes and share them with friends or people we fancied.

CDs were excellent too, but they were hard to love. In place of grooves, CDs had paths of binary-coded bumps so small only a laser beam could find them. What happens is the beam bounces off the bumps, and the sensor interprets the reflections at a rate of 44,000 times a second. I had first-hand experience of the CD, as one of my first full-time jobs was in a CD factory. During eternal night shifts (in fact, they lasted 12 hours), virtually every copy of the global supply of albums by ZZ Top, Take That and M People would pass through my hands in the post-print department. Perhaps this is where some of the enmity I've always felt for CDs came from. The plastic cases were nasty, the sleeves were hard to read, the discs themselves were hardy souls, yes, but they still went wrong, and once they had gone wrong there was no reasoning with them. To be honest, my most cherished CDs were some wonderful reissues from the early 2000s that had been made to look like exact replicas of their vinyl forebears – CD-sized card jackets, holding a paper inner sleeve, the discs themselves coloured vinyl black, with paper label-like artwork in the centre.

In the 1980s, a German engineer called Karlheinz Brandenburg started working on a 'digital jukebox', managing to create an MP3 of Suzanne Vega's 'Tom's Diner'. By the 2000s, with the emergence of MP3 players and iPods, the CD went into steep decline. Then suddenly, vinyl started to make another comeback. In 2016, vinyl sales reached their highest level since 1991. But even this return-from-the-brink narrative isn't so simple – cassettes have also made a tongue-in-cheek return via indie bands and hipsters, and the Recording Industry Association of America (RIAA) reported that shipments of compact discs rose from 31.6 million in 2020 to 46.6 million in 2021. That was the first time CD sales had gone up since 2004. So what on earth is going on?

I have a feeling it's down to several factors, one of which is that a large proportion of music lovers are realising that access to all music, all the time isn't all that it's cracked up to be. Most of us only end up streaming very limited portions of all that available music, and many of us still prefer to have physical objects as part of our fandom. Streaming companies, too, have attracted justified criticism for the royalties they pay to artists and the sound quality they offer, which isn't up to the exacting standards of many audiophiles.

The issue is sample rates. Sampling frequency refers to the number of times samples of the signal are taken per second during the analogue-to-digital conversion process. The audio sample rate gets all jumbled up with bit depth and bit rate. The bit rate is the sample frequency, which is multiplied by the bit depth and the number of channels. The bit depth is the number of bits per sample. And the important thing to remember is that all these bit depths and sample rates can quickly add up to rather large file sizes. A standard MP3 might have a bit rate of 320kbps, a CD would have 1,411kbps, whereas high-res audio would have more like 9,216kbps.

One of the great ironies of all this is that, in a sense, this striving for lossless, uncompressed audio is digital technology trying ever harder to recreate analogue sound. And we had analogue sound already. But still, the point is that the leading streaming services are catching up, streaming ever more high-res, lossless audio, pumping it through groaning networks and routers, stretching the limits of download speed and file size, swamping us with audio data our devices don't know what to do with. But as high-res audio moves mainstream, what will *that* mean for vinyl? Even if you wanted to get that onto vinyl, you couldn't – not in that much detail. It would be like me printing the entirety of *War and Peace* onto a single page. If I chose a small enough font, I could do it. It would all be there, every word, but it would be impossible to read. A cutting lathe can only cut a certain amount of frequency content in vinyl. And as an Abbey Road engineer told *What Hi-Fi?* in 2017, at around '20-something kHz', it just disappears.

The conclusion surely is that format doesn't matter. Music matters. We now have loads of different ways to listen to music, all of which can be made to sound awesome. When I was back at that CD factory in my teens, I was still something of a vinyl snob. Then something happened there that changed all that, something that still gives me goosebumps when I think about it, and still shapes how I feel about music and how I feel about formats.

The factory where I worked made a lot of cheap dance compilations, the kinds of discs that were put out by Cookie Jar records. Dance music of that era wasn't my bag, but it particularly wasn't my bag when it was presented in a seemingly endless procession of subpar megamixes. I just didn't get it. Why, when you're doing such a tedious repetitive job, for 12 hours straight, with no natural light, would you choose to listen to this 90-minute barrage of boredom?

I was a very low-status figure in the factory. There was no way *I* would be allowed to choose what we listened to, so I just had to endure it. And I can still hear it now. The bass drum, drilling into my skull ... Boots cats boots cats boots cats boots cats boots cats ... Strip lights, dry recycled air ... boots cats boots cats boots cats boots cats boots cats ... Sore feet, the second hand of the giant clock barely moving ... boots cats boots cats boots cats boots cats boots cats ... Warm plastic CDs, sweat-inducing latex gloves, still only 3 a.m. ... boots cats boots cats boots cats ... Finally, finally, the CD would come to an end, someone would stroll over to the CD player. Perhaps this time? Perhaps something else, an album maybe? Wait a minute ... Nope. Just another extended megamix of Haddaway hits. And the whole ghastly thing would start again.

One night, a woman who worked in the post-print department – essentially the next room, where you got to sit down as you worked – stood up from her workstation and walked slowly towards the pile of factory-second and off-cut mis-print CDs that collected near the stereo. She spent some time looking through the discs. Eventually, she found one, popped the lid, placed the disc in the player, closed it again, pressed play and walked back to her seat. In the moments of silence before the music started, I had very low hopes. I waited gloomily for the boots and cats to make their inevitable return.

Imagine that moment when instead of boots or cats ... came The Byrds singing 'Mr. Tambourine Man'. It was like opening a window. It was like water in a desert. It was like an actual angel coming down and sitting on my head. Tunes! *Harmonies*! Chiming guitars! And the thing is, that was no crisp, heavyweight, 180g vinyl, or lossless hi-res audio. That was a cheap, crappy, by-the-numbers, standard best-of 1960s compilation, played on a cheap portable stereo in the corner of a room. Yet it had me on my knees.

I didn't stop collecting records. And to be honest, I retained some of my vinyl snobbery for a while after that. But that moment was the seed of a new understanding, a moment from which I came to realise it wasn't the format that was important, it wasn't how you listened to anything. It was *what* you were listening to, and how what you listened to made you feel. And it all came from the groove. The groove not only changed how we listen, it changed what we listened to. And, through its fragile, narrow plastic walls, it delivered us here: to a place where we can play what we like, whenever we like.

Miscellany of the Groove

This isn't a comprehensive directory, dictionary, glossary of the early recording industry – that could fill several volumes. These are interesting labels, records, players, institutions, websites, oddities, sounds, one-offs, digressions and rabbit holes that I came across during my research. I'm particularly indebted to Michael Thosmas, whose exhaustive survey of 78rpm record labels in Britain formed the basis of many entries. I've also tried to give space to some companies, brands and gizmos that haven't featured elsewhere in this book.

78rpm Club
The website https://78rpm.club includes a survey of 78-producing record companies and labels from across the world.

'Abide with Me', 1920
Searching by the above should lead you to the first known electrically recorded music, captured at the Memorial for the Unknown Warrior, Westminster Abbey, November 1920.

Aco
A British record label active in the 1920s. You can explore an Aco catalogue from 1924 via the British Library's *Sounds*: https://sounds.bl.uk/Sound-recording-history/Early-record-catalogues.

Actuelle
Actuelles appeared in Britain in September 1921 and lasted until December 1928. The records used English, French and American Pathé masters throughout.

Amberol

The original Edison Amberols arrived in 1908. These were essentially the same as the Gold Moulded predecessors, but with double the number of grooves, so double the playing time. Machines that could play standard Gold Moulded cylinders couldn't manage the new finer grooves. The feed-screw mechanism kept the reproducing stylus directly aligned with the cylinder's grooves, meaning the new four-minute tightly packed grooves simply couldn't be played. Edison subsequently manufactured a machine that could play both by toggling between two different styli.

American Odeon

Made in the US and imported into Britain from 1905, these were advertised as 'Blue Odeon Duplex Records' as they were manufactured from a distinctive dark blue shellac. The records were double-sided, about 10½ inches in diameter and sold for 5s.

Ariel

Ariel Grand Records were produced for Messrs J. G. Graves of Sheffield, England, who sold them on a mail-order basis. These were available from 1910 until 1938, an amazing length of time for this kind of in-store label. The masters came from many sources over the years, including Beka, Favourite, Grammavox and Jumbo.

Arrow

This was an early British label pressed by the Carl Lindström group using Beka masters. It was a cheaper line, introduced as a result of the cut-price Cinch and Phoenix records from the Gramophone and Columbia stables, which themselves were a response to Lindström's Scala and Coliseum records.

Aspir

This French-manufactured record ('Aspir' being an anagram of 'Paris') was first available in the UK in May 1909. They were vertical-cut with an etched-and-filled 'label' similar to Pathé discs of the time. The records were sold from Victoria Street, London, the trading address of George Davies, who had been the agent for the French-made Phoebus and Phono Disc records in 1908.

Associated Glee Clubs of America, 1925
Columbia 9084 is one of the first electrical recordings issued on disc, and I think it's worth tracking down. It was recorded at the Metropolitan Opera House in New York on 31 March 1925. One side has an amazing recording of 850 members of the Associated Glee Clubs of America singing 'John Peel', the flip has the audience joining in on 'Adeste Fideles'.

Association for Recorded Sound Collections
The Association's 'Echoes of History' (www.arsc-audio.org) has a number of dissections of rare early audio. These include an early weight-loss exercise record from c.1920 and a 'snapshot in sound' made at the corner of Illinois and Washington Streets in Indianapolis in 1932. Recommended articles include Mason Vander Lugt's four-part series on the recording industry of the 1890s.

Autograph Records
This American label issued records of Jesse Crawford playing the Wurlitzer pipe organ in the Chicago Theatre using Orlando Marsh's own electrical disc recording system in 1924.

Auto Record
An 8-inch single-sided record manufactured by Beka Records in Germany, first advertised in the UK trade magazines in July 1905 as 'the cheapest record in the world' (sold for 6d). They were 'unbreakable', pressed on card, with a hardened playing surface.

Auxetophone
The Auxetophone was an early attempt to make horns louder for public events. These were compressed-air amplifiers. At the end of

the tonearm was an air-control valve, which opened and closed, sending pulsations of compressed air to the horn. Air pressure was provided by a ⅙ horsepower electric motor. There was no volume control and by all accounts it was unpleasantly loud unless heard from some distance. A 1906 advertisement for a 'New Pneumatic Victor' described the Auxetophone as producing a 'magnificent volume' that could fill the largest hall, theatre or church.

Babygram

One of the smaller 78rpm records ever produced, Babygrams were just 4 inches in diameter and single-sided, made from a vinyl-like material by British Homophone for Pippin Toys in the mid-1960s. There was a corresponding gramophone to play them on called 'The Babygram'.

Beka

Beka records were originally the product of Bumb & König, a German company involved with the German side of International Zonophone. When Gramophone bought Zonophone in June 1903, Bumb & König decided to introduce their own record label and named it after the company's initials (Beka is pronounced 'Bay-Kah'). In 1905 they sent recording engineers to London to record in Islington. By the following year they were selling 7- and 8-inch Beka records, the 10-inch Beka Grand and 11-inch Beka Symphonie. The 12-inch records Beka Meister in 1910 sold for 6s 6d each. The company eventually moved to 77 City Road, a four-storey building with the recording rooms on the top floor. In August 1910, Beka was bought out by Carl Lindström. In late 1912, a British factory was being built for manufacturing Lindström-owned records, all of which had previously been manufactured in Germany. It was called The Mead Works and was in Gas House Lane, Hertford. During the war the Beka name was dropped due to its German associations. One of its UK directors, Otto Heinemann, would found Okeh Records in America.

Berliner 5-inch records
The first disc records sold in America, from October 1894, were Berliner's first-generation 7-inch Gramophone records. However, the first sold anywhere in the world came before that, from the German toy maker Kämmer & Reinhardt in Waltershausen. The company made and marketed numerous Berliner discs that worked with a hand-operated cardboard-horned toy gramophone. This device was available from 1889 to 1892. The players rotated 5-inch records at 100 to 150 rpm. The records were made of gutta-percha, a kind of vulcanised natural rubber, and pressed from metallic matrixes. The label on the reverse of the records included the title and the words spoken on the record, like an early lyric sheet, to help listeners interpret the imperfect sounds. And the wonderful thing about these oldest disc records is that they are mostly singing or talking with no instrumental accompaniment, and they are nearly all Emile Berliner himself.

Berliner Lever Wind
A rare and short-lived machine from the early days of the Berliner disc, which used side-to-side lever power rather than the more usual rotating crank.

Berliner Style 5
This is the name of the 'Trademark' Berliner machine that would become part of the famous HMV logo. It was the model that artist Francis Barraud used to obscure the original 'Edison Bell' phonograph in his famous rehashed painting of Nipper the dog.

Biophone
A cumbersome, unusual and unsuccessful attachment invented in 1906 by Louis Devineau of Cleveland, Ohio. The idea was for a gizmo that could adapt a cylinder phonograph to play disc records. The mandrel of the cylinder would drive a heavy spoked turntable with weights at the edges, which acted as a flywheel to

help steady the speed. A tonearm and a sound box attached to the cylinder reproducer. A Biophone can be seen in the background of the 1964 movie *My Fair Lady*.

Black Diamond Band
One of the most popular British recording bands from about 1900 to the early 1930s – their last catalogue appearance came in 1932. Among the band's conductors were Eli Hudson and George W. Byng. The majority of their recordings were made for the Zonophone Company, but a few appeared on the Gramophone Concert, Monarch and His Master's Voice labels.

Blue Amberols
Introduced in 1912, these were the last incarnation of Edison's cylinder. While they represent some of the finest fidelity of the acoustic era, played for around four and a half minutes and were tough as all hell, for collectors they do suffer from shrinkage and deformation over time.

Bluebird Records
Started out as a low-cost RCA Victor subsidiary in 1932.

Bob
A Scottish record company, based in Glasgow, whose distinctive label design reproduced the reverse side of a shilling piece.

Bon Marche
These records date from the mid- to late 1920s and were manufactured by Crystalate in England for export to Australia for sale in the Bon Marche department stores.

Boots
In the early 1920s, the UK chemists started selling own-label records in a unique 6½-inch format. The records were recorded and manufactured by Vocalion around 1923.

British Homophone/Homophon/Homochord
A record-pressing company formed in 1921. They had agreements to use masters of the Homophon Company of Berlin, which had been making and distributing records under the Homophon banner since 1906. The German Homophon Company started exporting Homochord records in 1913. The British Homophone Company also produced their own masters and took complete control of pressing by the late 1920s. The company was eventually purchased by Crystalate, and the label finally disappeared in 1934.

British Library
British Library Sounds (https://sounds.bl.uk/Sound-recording-history) has historic interviews and audio snippets telling the story of recorded sound, as well as sections on archaic recording equipment.

Brown wax cylinders
This is the informal name given to the first generation of commercial Edison cylinders. After Edison and Graphophone entered a period of patent-sharing, Graphophone's wax-coated cardboard tubes were abandoned, and Edison's all-wax cylinders became the first standard format. The very earliest tended to be white, derived from a blend of plant and animal waxes. However, these weren't uniform enough, so they soon switched to a standardised metallic soap composite. There was no room for any identifying marks on the cylinder itself, so the name of the song or artist was normally included on a little slip of paper that went into the box that held the cylinders. They were made of a relatively soft wax, which wore out after they were played a few dozen times, but they could also be shaved and re-recorded. Various sizes of cylinder appeared, but the late 1880s 'standard' were about 4 inches (10cm) long, 2¼ inches (5.7cm) in diameter, and played about two minutes of sound. Cylinders were sold in cardboard tubes with cardboard caps on each end. The earliest

soft-wax cylinders came swathed in a separate length of cotton batting. Later hard-wax cylinders were sold in boxes with a cotton lining. Celluloid cylinders were sold in unlined boxes.

Brunswick
This American label was formed in 1916 by Brunswick-Balke-Collender, a company based in Dubuque, Iowa. The name came to Britain in 1923 and was bought by Decca in the 1930s.

Brunswick Panatrope
The first fully electric record player, with electromagnetic pick-ups in place of the old acoustic sound box.

Capitol
Founded by Johnny Mercer, Buddy DeSylva and Glenn E. Wallichs, Capitol was the first to support all three recording speeds of 78, 45 and 33⅓ rpm.

Catalogues
Via British Library Sounds (https://sounds.bl.uk/Sound-recording-history/Early-record-catalogues) you can read record catalogues from the acoustic era, from the likes of Columbia, Edison, Gramophone and His Master's Voice, as well as smaller names such as Aco and Homochord, and overseas companies such as Odeon and Pathé. An Edison Bell catalogue from 1898 reads: 'We should esteem it a favour if customers, when ordering records, would name a few extra ones as "second choice". Owing to large demand for some records, we are occasionally cleared out of them ...'

Century Old Sounds
The website www.centuryoldsounds.com features many early recordings from a personal collection of discs and cylinders, from around 1895 to the Big Band era.

Chess Records
Founded in 1950 in Chicago, Chess specialised in blues, rhythm and blues, and later soul, gospel and jazz. Subsidiaries included Checker, Argo and Cadet.

Cinch records
Cinch records (c.1913) marked the start of a cut-price war between the major record companies in Britain trying to regain ground lost to cheaper newcomers such as Coliseum and Scala records. Cinch records were manufactured by the Gramophone Company, often using existing Zonophone matrices.

The City of London Phonograph and Gramophone Society
Formed in 1919, as the London Edison Society, this is the senior recorded music society in the world. Current patrons include Oliver Berliner (grandson of the inventor of the gramophone) and Simon Blumlein (son of Alan Blumlein, the inventor of stereo recording). The Sound Library (www.clpgs.org.uk) includes recordings by Clarion, Sterling, Pathé, Columbia and Edison Bell. You can also listen to early cylinders as well as later four-minute Amberols. I particularly recommend the 'Ragtime Goblin Man' sung by Arthur Collins and Byron G. Harlan, recorded in June 1912. There's also the Blue Amberol 'Everybody Loves a "Jass" Band', sung by Arthur Fields in July 1917. According to the description, the word 'jass' was too susceptible to a deleting of the J on posters, so was quickly respelled with two Zs. You can read back-issues of the society journal *Hillandale News* via https://archive.org/details/thehillandalenews.

Clarion
English label Clarion made cylindrical records as well as two types of disc record.

Coliseum
Appeared in 1912 offering popular music at a budget price and continued until 1927. The records were manufactured in Clerkenwell Road, London, using matrices from Vocalion and Gennett.

Columbia
Obviously, we've covered Columbia elsewhere in this book, but it's just interesting to note that Columbia discs first appeared in Britain in 1902. By then the American Columbia Phonograph Company had set up a branch office in London, initially importing wax cylinders and Climax records. Then, in 1906 they started manufacturing, opening a factory in Bendon Valley, Garratt Lane, Wandsworth.

Compatible Discrete 4
Also known as Quadradisc, this was a discrete four-channel quadraphonic system created by JVC and RCA in 1971.

Concert cylinders
These were larger-diameter brown-wax cylinders brought out in 1898 (Columbia's were called 'Graphophone Grand' records, Edison's were 'Concert Records'). These were 5 inches in diameter and could deliver a louder recording. First recorded at 120rpm, then 144rpm in late 1899, and finally 160rpm in early 1902. Edison continued making Concert cylinders until late 1911.

Crystalate
Crystalate Manufacturing Company was a British plastics and later electronic components manufacturing company that operated between 1901 and 1990. Based in Tonbridge, Kent, it was producing record matrices for more than 20 other companies by 1906.

Crystal Palace, 29 June 1888
A chorus of 4,000 voices 'recorded with phonograph over 100 yards away' by Colonel Gouraud at Crystal Palace, London. You

can hear it here: home.nps.gov/edis/learn/photosmultimedia/very-early-recorded-sound.htm.

Curry's
Curry's were bicycle manufacturers who started selling records and gramophones too. They also branched out into having their own record label.

Danceland
A UK series of records that first appeared around 1948 – the result of differences between the Association of Ballrooms and the record companies regarding licences to use gramophone records in dance halls.

Decca
Decca had been manufacturing gramophones in Britain since 1914 but entered the record-making business in 1929. They took over various labels following the takeover of Crystalate in 1937.

DeLuxe
Label formed in 1944 by brothers David and Julius Braun in Linden, New Jersey. Rock 'n' roll landmark 'Good Rocking Tonight' by Roy Brown was released in 1947.

Delyse
The Delyse Recording Company of Wallington Road, London, was owned by Isabella Wallach, the great-niece of Fred Gaisberg.

Diamond
As well as Edison's first discs, this was the name of a label that made 10½-inch vertical-cut discs produced by Pathé from 1915 for the Diamond Disc Record Company Ltd of 81 City Road, London. Like Pathé records of the period, it had no paper label to begin with – instead the words were etched in light blue.

Dictabelt

The final cylinder-type sound carriers were 'dictabelts', thin, flexible cylinders of a kind of red, blue or purple vinyl that was stretched between two rollers and set in rotation in a dictaphone machine. Variations of the machine were used to record the telephone conversations of President Kennedy, and Agatha Christie began using dictabelt towards the end of her writing career. The format was also used to record the landmark Rivonia trial, in which Nelson Mandela, Walter Sisulu, Govan Mbeki, Raymond Mhlaba, Elias Motsoaledi, Denis Goldberg, Andrew Mlangeni and Ahmed Kathrada were sentenced to life imprisonment. The proceedings were recorded on 591 dictabelts, which have been inscribed on UNESCO's Memory of the World register. You can listen to some excerpts here: https://blogs.bl.uk/sound-and-vision/2013/12/rescuing-the-rivonia-trial-recordings.html.

Dimestore labels

A group of American low-cost labels that included Perfect, Oriole, Banner, Melotone and Timely Tunes.

'Down Hearted Blues'

In 1923, Bessie Smith's recording of 'Down Hearted Blues' sold 750,000 copies for Columbia.

Duophone

This rather eccentric label started out using shellac, before aviator, inventor and publisher Noel Pemberton-Billing devised an 'unbreakable' disc using a layered paper core covered with a sort of plasticised rubber. They also produced 'lightweight and unbreakable' Fetherflex records.

Durium

This was a mid-1930s UK equivalent to the US 'hit-of-the-week' records, available weekly from newsstands. Made of cardboard with a thin celluloid coating, they used fine grooves to give space for two tunes per side. This was also the name of an Italian record label active from 1935 to 1989.

Dynaflex
During the early 1970s vinyl slowly became thinner – RCA Victor called their new, thinner product Dynaflex.

Dynagroove
Another new recording system from RCA Victor, designed to reduce surface noise.

Edison Bell Phonograph Company
Formed in 1892, this company began life as the holder of distribution rights in Great Britain. There was tension with the company parent (American Edison), who kept selling American-made machines in the UK territory. They also had skirmishes with a sewing-machine salesman from Manchester named James E. Hough, who formed the company Edisonia. These eventually merged to become the Edison Bell Consolidated Phonograph Company. When the Bell-Painter patents expired, they moved into both distribution and manufacture of machines and records, eventually becoming one of the major global producers of phonograph cylinders, alongside National, Pathé and Sterling. Edison Bell's cylinders were the dominant brand in Britain from the early 1890s until 1903, when the patents expired. In 1908 they produced the Bell – their first disc-shaped records. (They also revived this name in the 1920s for their children's label.)

Edison Bell Picturegram
An obscure, mid-1920s portable-but-not-really machine. It came with a paper roll of pictures that would accompany fairy stories being played on the record.

Edison Class M
Four stethoscope-like tubes attached to an Edison Class M electric phonograph fitted inside an oak cabinet became the first jukebox. This was installed in San Francisco's Palais Royale Saloon in 1889.

Edison Phonograph Monthly
Fascinating American trade magazine that ran from 1903 to 1916. You can read copies via https://mediahistoryproject.org.

Edison Talking Dolls
You can listen to digitised audio from eight Edison Talking Doll recordings, made between 1888 and 1890 (www.nps.gov/edis/learn/photosmultimedia/hear-edison-talking-doll-sound-recordings.htm). These come from the Thomas Edison National Historical Park collections.

Eiffel
In September 1889, during a two-month trip to Europe, Edison was invited to a grand luncheon at the Restaurant Brébant on the first level of the Eiffel Tower. After the meal the celebrants were invited up to the very top of the tower, where Gustave Eiffel kept an apartment, and Edison presented Eiffel with a Class M phonograph.

EMI Music Archives
The archive trust (www.emiarchivetrust.org) looks after about 18,000 'Berliners' – first-generation commercial disc recordings made towards the end of the nineteenth century, many collected by Fred Gaisberg during his 'Grand Tours' across Europe, Asia and the Far East. Staff recently spent two years scanning the often hard-to-read labels as a first step towards cataloguing the discs.

Encore
This was a pre-First World War Beka product, pressed in Germany (where it was marketed under the name of 'Veni-Vidi-Vici'). The record's unique feature was that it offered two titles on each side, rather like the later 'four-in-one' discs. The difference was, though, that Encore discs still only played for around the normal three minutes, so they achieved the multi-track feet by the simple trick of only including very short numbers.

Famous
One of the many record labels made for J. Blum & Co of 220 Old Street, London. Famous records first appeared in September 1912 and were initially pressed in Saxony, Germany, by Kalliope. In 1913, Blum fell out with Kalliope due to them pressing records for other clients but using Blum's exclusive matrices.

Favorite
A major record manufacturer in the early years, this German company began exporting to England in 1908. From the First World War their records were pressed at the Mead Works in Hertfordshire.

Feaster, Patrick
Patrick Feaster's 'explorations in historical media' (https://griffonagedotcom.wordpress.com) delve into the earliest recorded sounds, the cultural impact of the phonograph, as well as material on telephones, early movies, other media and ciphers. Recommended reading includes an article looking at the phonograph for secretarial dictation; a 'phonographic funeral' from 1895; a piece on 'Lopograph Records' from 1873 to 1878; and some reconstructed 'stereo' from 1901 – Berthold Laufer recorded Chinese music and drama using two phonographs at the same time. There's also a forensic study of a phonographic time capsule that was placed beneath

the cornerstone of the New York World Building in 1889 ... and appears to have been switched!

Filmophone
This was an attempt to come up with a new type of record, in a kind of thin, film-like format. Issued in the UK, they were in fact a little too flexible and tended to quickly become unplayable.

First Sounds
The firstsounds.org collective seeks to unlock and make available humanity's earliest sound recordings. This is a good place to find digitised phonautograms and listen in on the Volta Lab experiments of the 1880s.

Fontana
Dating from the 1950s, Fontana records were a product of Philips Electrical Ltd.

Four-in-One
This was a British Homophone product, and a genuine attempt to give buyers more value for their money. The records include four full-length dance band tunes on every record, each side playing for at least six minutes, which was long-play for the time (early 1930s).

Franklin Institute, Philadelphia
This is where Emile Berliner demonstrated his first gramophone on 16 May 1888, using a flat 7-inch disc with lateral-cut grooves.

Gaelfonn
A Glasgow-based record label from the 1950s, Gaelfonn concentrated on Gaelic (Scottish) material. The label was founded by Murdo Ferguson, a Gaelic speaker from the Isle of Lewis.

Gennett
This Indiana record company began making lateral-cut records and was promptly sued by Victor in 1919. Together with a small group of American labels, it successfully defended the claim that lateral-cut was now in the public domain.

Gold Moulded
Edison Gold Moulded Cylinders were made by creating a metal mould from a wax master – a brown wax blank could then be put inside the mould, heated, and as it expanded, the grooves would be pressed into it. These were played at 160rpm and ran for about two minutes. By 1902, Columbia was also putting out its own gold moulded-style cylinders. For some years both companies put out releases on both brown wax and moulded black wax cylinders.

Goodson
Another odd flexible format, which appeared in the late 1920s and early 1930s. Named after the company's owner, Jack Goodson, they were made from a flexible white material known as Rhodoid.

Gramophone
Although the word 'gramophone' became a term for a record player, it was of course originally the trade name of the Gramophone Company. For their own records they used the 'recording angel' trademark in the early 1900s, before eventually switching to Nipper and 'His Master's Voice'.

Grand Opera Cylinders
These were a special series of high-brow Gold Moulded Cylinders, later continued in the four-minute Amberol format.

Graphophone B
Graphophone B was perhaps the machine that really launched the phonograph into the home: a simply constructed machine, costing $10, that played cylinders that lasted two minutes.

Great Scott
A label set up by John Drummond (the 15th Baron Strange) at his home at Megginch Castle, Perthshire, Scotland, in late 1933. Drummond purchased the recording equipment from Edison Bell and made records of local artists and Scottish music.

Grey Gull
Record company and label founded in Boston, Massachusetts, in 1919, which started life selling high-quality vertical-cut discs.

Guardsman
These discs first appeared in May 1914. This was the result of a disagreement and a court order imposed on William Barraud to stop him using the 'Invicta' name for his records. In 1922, Vocalion took over the label.

Harvard Disc Record
This label was manufactured in the US by the American Graphophone Company (Columbia) between 1905 and 1907. The name was first applied to inexpensive external-horn phonographs that were sold through Sears Roebuck, and the record label was produced to match the phonograph.

His Master's Voice (HMV)
Probably the most famous name in record production in the world. In the UK it became the trade name for the Gramophone Company in 1910. HMV records was the premium line, costing more and with the top-billing artists.

H.M.S. Pinafore
In 1907, British firm Sterling brought out a 10-cylinder recording of Gilbert and Sullivan's *H.M.S. Pinafore*. This was a very early attempt to produce a kind of full album collection of a popular show.

Holzweissig Hymnophone
Although we might think of Victrola as the original internal horn machine, it wasn't. The modestly sized 1904 Holzweissig Hymnophone came first.

Hudson
A British record company that put out the first recording by Vera Lynn.

Humming Bird Records
A mid-twentieth-century record label based in Waco, Texas, that focused on local and regional artists including Cajun music.

Imperial
Produced in England by the Crystalate Gramophone Company from 1922 until 1934, this label is notable for putting out some 78rpm 3½-inch promo discs.

Indestructible Records
The Indestructible Phonograph Co. of Albany, New York, produced moulded celluloid cylinders beginning in 1907. The company made both two- and four-minute cylinders, and the repertoire was similar to that of Edison and Columbia cylinders. The cylinders had a thick cardboard core and metal rim ends to keep them rigid.

IRENE
This is the non-contact, optical-scanning project that has digitised and unlocked audio from fragile and broken grooved

audio formats. The website (irene.lbl.gov) includes a page with all the audio used for a presentation given by Carl Haber in June 2021, including 'Au clair de la lune' from 1860, an Edison tin-foil recording from 1878, and an aged and damaged wax cylinder from an old fortune-teller machine. 'You will soon go to a ball or a large gathering and meet a new friend …'

Jumbo
The first Jumbo records appeared in Britain in 1908, as a cheap equivalent to Fonotipia. The records were initially made in Germany but subsequently pressed by Crystalate and later the Mead Works in Hertfordshire.

Kämmer & Reinhard Berliner
The first commercially produced Berliner disc player was a 'toy' machine produced by Kämmer & Reinhard in Waltershausen, Germany, in 1889.

Kapp
This was an American label founded in 1953 by Dave Kapp, brother of American Decca founder Jack Kapp.

Lambert cylinders
Thomas Lambert developed a successful method of mass-duplicating 'indestructible' cylinders. The distinctive 'Pink Lamberts' were manufactured by the Lambert Company of Chicago from 1900. These were celluloid and came in a variety of shades of pink as well as purple, brown and black. While Edison would eventually put Lambert out of business, some of the cylinders he recorded in Chicago are believed to be the earliest surviving recordings of Yiddish songs.

Language courses
There were lots of different language courses. The 'Foylophone' series, for example, was published by London bookseller Foyles.

Hugophone was a Columbia-manufactured series of colour-coded discs – French was red, German black or dark blue, Italian green, *etc.* This followed the pattern set by the Linguaphone Language courses, started by Clare Rees and first issued on cylinder.

Levaphone
Levy's of 139 Whitechapel High Street, London, were famous among jazz record collectors for making many rare American jazz items available both through their own labels and their importing service.

Library music labels
Many labels specialised in producing 'light music' that could be used by radio, film and TV as background. In Britain, an example was Conroy, who produced vinyl discs for the Berry Music Company in the late 1950s and into the 1960s. KP Music, dating from the late 1950s, were produced for the Keith Prowse Publishers, while Bosworth records were introduced in early 1937. These were initially recorded and pressed by Decca and subsequently by EMI.

Library of Congress JukeBox
This is the name of a collection of digitised audio from the Library of Congress (www.loc.gov/collections/national-jukebox/about-this-collection/). It includes more than 10,000 recordings made by the Victor Talking Machine Company between 1901 and 1925.

Lioret Cylinders
French clockmaker and inventor Henri Lioret developed a talking doll in 1893 using a small celluloid cylinder. He went on to develop cylinders that had a brass tube with spokes covered by a moulded celluloid sleeve carrying the grooves. This makes them the first sound cylinders to use the more

durable celluloid, which Edison would eventually adopt for his Blue Amberols. By the late 1890s he was making several sizes of musical cylinders designed for clockwork phonographs called Lioretgraphs.

Lomax Digital Archive
Another place where you can access hours of unique audio/visual material, this time compiled across seven decades by folklorist Alan Lomax and his father John A. Lomax. This includes instant disc recordings from the Library of Congress's Archive of Folk Song between 1933 and 1942 (https://archive.culturalequity.org).

London
This famous label was produced by Decca initially for export to America.

Longanote
Longanote records date from the very early 1930s and are another example of a lesser-known attempt to create long-playing discs in the UK. These were capable of up to 15 minutes of playing time on each side, made possible by recording with a constant linear speed – a specially designed motor that spun more quickly as the tonearm reached the centre. By March 1932, these were taken over by Filmophone, by which time they were being billed as either 10-inch discs with 12 minutes of playing time per side, or 12-inch records with 18 minutes each side. By December 1932, Longanotes were being made by Edison Bell.

Magnavox
Edwin S. Pridham and Peter L. Jensen in Napa, California, invented a moving-coil loudspeaker they called the 'Magnavox' in 1911. It was used by Woodrow Wilson in San Diego in 1919.

Marathon
This was the name of another early long-playing record, available from 1912 to 1915. It was the invention of Percy J. Packman, who hit on the idea of a V-shaped vertical-cut groove that allowed a very fine pitch to be used while still maintaining 80rpm. This gave five minutes of playing time on a 10-inch disc and eight minutes on a 12-inch.

Marconi
These discs began as flexible, 'unbreakable' single-sided records, with a card or thick paper core, coated in a hardened plastic surface on which the groove was pressed, manufactured by the American Graphophone Company using Columbia masters.

Matrix
Once a wax master record was cut, it would be electrotyped into a metallic matrix.

May-Fair
May-Fair records were available between 1931 and 1933 and were not sold in the usual manner, instead exchanged for Ardath cigarette coupons. The records were pressed mainly by Brunswick and Piccadilly, and later by Edison Bell and Decca.

Mechanical recording
An alternative (and in some ways more accurate) term for acoustic recording.

Memorial Record
This was the label of a special 12-inch record made by Columbia of the funeral service of the Unknown Soldier. This took place at Westminster Abbey on 11 November 1920 and is a first issued electrical recording, taken by Guest and Merriman. The record original was sold for 7s 6d, the money going towards the Westminster Abbey restoration fund.

Mercury
Mercury Records engineer C. Robert Fine pioneered a monaural recording technique in 1951, branded the 'Mercury Living Presence' records.

Mikiphone
This was a tiny Swiss-made pocket record player the size of a large pocket watch, which could manage discs of up to 10 inches.

Milton Brown and His Musical Brownies
If you'd like to hear some pretty wild electric guitar, considering it was recorded in the 1930s, search out some of Bob Dunn's amplified lap steel. A particularly good example is 'Taking Off'.

Montross
In early 1896, Berliner contracted with Eldridge Johnson to develop an improved spring motor for the gramophone. Johnson turned to machinist Levi Montross – they filled the first order of 200 machines that Berliner requested by 10 August. They were sold for $4 each wholesale.

MSS
This was one of the major UK manufacturers of acetate (lacquer) discs used by private recording studios. The name stood for Marguerite Sounds Studios and later Master Sound Studios. It was started by Cecil Watts, who developed the disc after witnessing a 1920s recording session and seeing the enormous effort and time spent using thick wax masters. Later he invented the 'dust bug' – a clever little brush attachment that sat on turntables, gently clearing a path in the dust before the stylus did its stuff.

Musée de la Voix
On Christmas Eve 1907 a time capsule was walled up in a basement storeroom of the Paris Opéra. The capsule actually

comprised two cylindrical lead urns holding 24 disc recordings of the great opera stars and instrumentalists of the day. Each was carefully wrapped in asbestos-covered cloth and separated by sheets of glass, and were not to be opened for 100 years. It was called the Musée de la Voix, and a second 'burial of the voices' took place in 1912. This was the brainchild of Alfred Clark, a music-loving New Yorker who had previously worked with Edison, making early short movies at his legendary Black Maria studio, before managing the Compagnie française du Gramophone. The vault had Gramophone Company recordings by Franz, Caruso, Amato, Adelina Patti, Emma Calvé, Bessie Abbott, Nellie Melba, bass Feodor Chaliapin, sopranos Marcella Sembrich, Geraldine Farrar and Luisa Tetrazzini, a piano piece by Ignacy Jan Paderewski, another by violinist Jan Kubelík, and a piece played by the Band of the Coldstream Guards. The second batch also came with a hand-cranked gramophone and a supply of needles. During building works in 1889 it was discovered that the archive had been raided. One of the 1912 urns was empty, the gramophone was gone. The remainder of the archive was immediately transferred to the Bibliothèque nationale de France. And at the end of 2007, the surviving archive was opened. As the urns included printed records of what was supposed to be in each cache, they were able to piece together the missing contents.

National Recording Registry

US Congress established the National Recording Registry for the purpose of maintaining and preserving sound recordings that are culturally, historically or aesthetically significant. You can read the list of entrants here: www.loc.gov/programs/national-recording-preservation-board/recording-registry/complete-national-recording-registry-listing/.

Odeon

A major international label during most of the 78rpm period, founded in 1903 by Max Straus and Heinrich Zuntz of the

International Talking Machine Company. Odeon became a part of Swedish inventor Carl Lindström's empire, which also owned Beka, Parlophone, Fonotipia, Lyrophon, Homophon and other labels. Lindström was eventually acquired by the English Columbia Graphophone Company in 1926, which, when it merged with Electrola, HMV and other labels, turned into EMI. The original Odeon discs on sale in the UK were double-sided 7½-inch and 10¾-inch, costing 2s 6d and 5s respectively. This was at a time when most records were still single-faced.

Okeh
American record company founded in 1918 by German emigrant Otto Heinemann.

Olympic
These discs were produced for Levy's Phono and Cycle Stores of High Street, Whitechapel, London.

Overseas Recorded Broadcasting Service
Set up in the Second World War to provide recorded entertainment for forces stationed in Britain.

Packard-Bell PhonOcord
These were a series of heavily advertised all-in-one American machines from the mid-1940s that included radio, phonograph and home recorder. Another brand of home recording was the UK Permarec discs, made by Musikon Ltd from late 1933. The records were a very thin metal sheet coated in a black substance. These could be recorded using Permarec's home recording device, then baked for two to three hours at 100°C (212°F)!

Parlophone
The introduction of Parlophone in Britain marked the resurrection of trading with Germany after the First World War.

The Lindström Company that owned the label used their old factory in Hertfordshire from the early 1920s, leasing it from the then-owners Columbia.

Paste-over labels
This was a way of rebranding and reselling old stock. So to take a totally random example: if you chance upon a disc with a 'Cameo records' label, which were sold by Gilbert's of Sheffield for 1s, you're in fact holding an old Regal record that had been re-labelled. Similarly, 'Crown Perfects' were in fact old single-faced Columbia records, sold by Selfridge's. Sometimes they'd only cover half of the label – removing the brand but leaving the track information. 'Ogden Smith Top Note' were over-labelled discs sold by a bicycle accessories dealer based in Cheapside, London.

Pathé
An international record company, label and producer of phonographs, based in France and active from the 1890s through to the 1930s. Their unique vertical-cut discs came to Britain in October 1906, by which time the firm had studios in London, Milan and Moscow. The size of early Pathés varied enormously – from 7 inch to 20 inch, and usually played at between 90 and 100rpm, with a wide U-shaped groove that started at the centre.

Pathé cylinders
In addition to standard 2¼-inch x 4-inch cylinders, Pathé also manufactured 3½-inch x 4-inch 'Salon' cylinders.

Philips
Philips Electrical Ltd was based in Eindhoven, the Netherlands. They started manufacturing records for the British market in 1953.

Phoenix

Launched in 1913, Phoenix was Columbia's answer to the cheap German imports that ravaged the record industry at the time. They cost 1s 1d, bore the usual mix of popular vocal, marches, ragtime, xylophone and banjo music, and were designed to challenge cheap incomers such as Coliseum and Scala.

Phonobase

This site (www.phonobase.org) has sound excerpts taken from early commercial cylinders and records made from 1888 to 1920, distributed in France and Europe mainly.

The Phonogram

A monthly trade magazine devoted to the 'science of sound and recording of speech', edited by Virginia H. McRae, a typewriter entrepreneur from Wilmington, North Carolina, who ran her own copying business in New York. In the first issue she wrote: 'The object ... is to familiarize the public with the good qualities of its namesake, to preserve the record of its growth while it moves forward to the achievement of the highest possible good, and to illustrate the part it performs in the work of human progress.' The magazine ran for three years, capturing the burgeoning industry during its 'brainstorm' period – where the first-generation regional companies were basically trying anything and everything to make money out of the new invention. It ran between 1891 and 1893, and you can read back-issues via https://archive.org. Please note, there was another magazine named *Phonogram*, published in 1900.

Phonograph Party

Listen to the original audio from the famous 'Phonograph Party' held by Colonel Gouraud at Little Menlo on 5 October 1888, painstakingly pieced together by Patrick Feaster: https://griffonagedotcom.wordpress.com/2017/04/05/the-phonograph-as-toastmaster-october-5-1888/.

The Phonoscope
You can read back-issues of the first independent magazine for the recording industry, set up by Russell 'Casey' Hunting in 1896 at https://archive.org/details/phonoscope13hunt.

Plaza
Plaza was British Homophone's leading entry for the 8-inch market. They first appeared in 1933 and had a 'strobe' design, like all the BH products of the period. To keep costs low, they usually had non-copyrighted material on one side.

Polly Portable
The Polly Portable was a portable disc gramophone from around the 1920s that came with a collapsible 'paper' horn as reproducer.

Polydor
This name was originally launched by the German Polyphon Company just after the First World War.

Pressing
The standard process of pressing records had five steps: master, matrix, mother, stamper, pressing. The 'mother' was a robust replica of the original positive wax master. This could then be used to electrotype more stampers for making more records.

Private recording companies
Specialist companies offering private recording services, which could either take the form of one-off direct-cut (acetate) discs, or pressings, cropped up all over the place. Atlas was based in Edinburgh; Glasgow had Barrie Hall Recording and Biggars; Belding & Bennett company were based in Wallington; GC Electrical were in Wardour Street; Bonham & Wilson were in Coventry, while the Eccles Recordings Services served west of Manchester. The

Liverpool company Kensington is most famous for making a disc of The Quarry Men – recorded by future members of The Beatles. Percy Phillips ran the recording studio and cutting service 18 Kensington, Liverpool, from 1955 to 1969. The first disc home recording device was put on the market by The Nicole Record Company in late 1904 as the 'Autorecordeon'. These were pre-grooved discs, the groove filled with a waxy compound. A couple of years later the short-lived Neophone brand marketed a similar device that cost 30s. When Neophone folded, their home recording stock was bought by Edison Bell, who rebadged the attachment the 'Edison Bell Eureka'. There's a survey of some home recording devices sold in Britain at: early78s.uk/disc/. A page on American home recording discs can be found here: www.phonozoic.net/recordio/index.htm.

Puratone
The Puratone Record Company's factory was the Arcadia Works, Church End, North Finchley. These were bespoke picture discs produced for advertising purposes in the mid-1930s.

Pye
The Pye Radio Company started issuing records in 1954 after acquiring the Nixa Record Company.

Radio
Edison Bell's entry into the lucrative 8-inch disc market came in 1928 with Radio records. Ex-music hall artist Harry Hudson was musical director and provided most of the dance music under a variety of pseudonyms.

Regal
Started as a budget label made by Columbia Records and introduced in early 1914, costing 1s 6d.

Reproducer/Sound Box
The sound reproduction of an acoustic phonograph/gramophone starts with the stylus or needle that is attached to the reproducer or sound box. A diaphragm is made to vibrate in a tube, the sound of the vibrations are passed to the horn, which then increases the volume depending on its length.

Rex
The 'King Of Records' first appeared in September 1933, produced by Crystalate and sold in Marks & Spencer stores across Britain.

'Rhapsody in Blue'
This Gershwin number gives us a useful compare-and-contrast for the switch between acoustic and electric recording. If you search by 'First Recording of Rhapsody in Blue, Paul Whiteman, 1924 version', you should find the original. Then if you look for 'Rhapsody in Blue, Paul Whiteman with George Gershwin, 1927 version', you should find the follow-up. Even allowing for the changes in arrangement, it's a marked difference that is easy to hear.

Royal Purple Amberol series
These were a lesser-known offshoot of the Blue Amberols – essentially identical, just dyed purple – that were filled with high-class classical and opera numbers.

Seeburg
One of the 'big four' coin-operated phonograph companies in the States. The Seeburg 'Select-O-Matic' jukebox had a row of linear slots from which it could play discs vertically clamped to a turntable. In 1959 it launched Seeburg 1000, which had its own line of 9-inch, 16rpm records.

Shellac
This has become the catch-all term for records of the era, although in reality it was just a common ingredient in a very

wide range of materials, other ingredients including carbon black and coal dust. Some discs from the 'shellac' era didn't use any shellac. Some were pressed in celluloid, for example, or had a layer of celluloid on some other core.

Specialty Records
Art Rupe started Juke Box Records in 1945, which would become Specialty Records in 1947, growing into one of the most important independent labels of rhythm & blues and early rock 'n' roll. Rupe passed away on 15 April 2022 at the age of 104.

Stereo
This playlist from the British Library includes audio from Alan Blumlein's actual experiments into two-channel recording: https://sounds.bl.uk/Sound-recording-history/Alan-Blumlein-recordings.

Sterling
Upon the expiration of Edison Bell's patents in England, various manufacturers began selling cylinders. Louis Sterling was an American who had moved to England and was employed by the British Zonophone. He resigned in 1904 and started Sterling in partnership with Russell Hunting, who became recording director of the company. The resulting Sterling cylinders were a little longer than standard cylinders.

Stroboscopes
These were circles of equally spaced lines on a contrasting background – usually black and white – printed around the edge of record labels. Electric lights are constantly switching on and off, too rapidly for us to see. When viewed under an electric light the lines appear to stand still when the turntable is rotating at the correct speed, as each time the light blinks off, the black lines on the stroboscope have rotated just one space.

Talking Machine World/Talking Machine News
These were two trade titles that you can read via https://archive.org or https://mediahistoryproject.org.

Tally Man
This was the term for types of labels/records that were sold door-to-door in the UK. The buyer had to sign a contract, agreeing to purchase a fixed number of records over a fixed period, and for this they would receive a 'free' gramophone on loan. The actual labels tended to have patriotic names like John Bull, Flag or Britannic. As an example, a customer had to agree to buy about 50 John Bull discs over the course of a year – costing 2s 6d each. The John Bulls dated from 1909, and used masters from Beka and Favorite. It is thought that the influx of cheap records killed off tally-man businesses.

They Might Be Giants
In 1996 the band They Might Be Giants recorded 'I Can Hear You' and three other songs, performed without electricity, on an 1898 Edison wax recording studio phonograph at the Edison National Historic Site in West Orange, New Jersey.

'Thirteen Women'
'Rock Around the Clock' by Bill Haley and His Comets was first issued as a B-side to 'Thirteen Women (and Only One Man in Town)'. The band recorded with Decca on 12 April 1954 in New York City. The guitar solo was played by Danny Cedrone, who died falling down the stairs 13 days after the session.

Thomas Edison National Historical Park
Preserves approximately 28,000 disc records and 11,000 cylinders. The 'Listen to' page has various curated playlists (home.nps.gov/edis/learn/photosmultimedia/the-recording-archives.htm). Under 'Experimental' you'll find some Diamond Discs recorded in the 1920s using a 125ft long recording horn.

The 'Very early' section has the Crystal Palace recording from 1888, 'the Lost Chord', an after-dinner toast from Little Menlo in London, and 'Around the World on the Phonograph' – believed to be the earliest recording of Thomas Edison's voice.

Tonearms
In 1903, 12-inch discs arrived. At the same time, the tonearm was introduced by Victor/Gramophone. This was a hinged metal tube that carried sound from sound box to horn. Because these new pivoting tonearms did all the moving across the disc as the stylus tracked the grooves, this meant that the horn could stay in the same place. This, in turn, freed up makers to produce heavier, louder horns with disc machines, and to experiment with putting them in new positions – including internal horns. In 1904 the distinctive 'Morning Glory' horns appeared – with petals of steel or brass. Later, this shape would be recreated with laminated wooden horns.

Top 40
Todd Storz of Omaha's KOWH created the Top 40 in 1949 after observing customers in a bar play the same jukebox selection over and over.

Transacord
This company was started by Peter Handford, who was an experienced sound engineer and sound effects expert. In 1953, with just a tape recorder and a 78rpm lacquer disc cutter, he offered private recordings and also recordings of amateur music events and competitions. He would cut the records himself, from his tape recording, unless more than 12 copies were required, in which case he would get records pressed by British Homophone. In 1954, he heard an American LP of railway sounds and decided to try something similar with British railways. His first records in this style were issued in 1955 on

78rpm discs. In 1958, after about half a dozen 78rpm issues (10-inch and 12-inch sizes), he switched to the LP and 45rpm format. Transacord was later taken over by Argo Records and subsequently became part of Decca.

Trusound
One of the earliest flexible picture discs, Trusound date from the early 1930s and are very scarce today, and when they do turn up are often unplayable. The Trusound recording studio was in a former church in London's St John's Wood.

Twentieth Century Talking Machine Record
Starting in 1905 the Columbia Phonograph Company released a short-lived series of 6-inch long Twentieth Century Talking Machine Record cylinders that played for three minutes rather than the usual two minutes.

Twin
This was the label under which the Gramophone Company launched its first double-sided discs in 1908.

Type N Graphophone
The 1895 Type N was one of the earliest spring-motored models, and the first to have a fixed mandrel for Edison-style cylinders.

UC Santa Barbara Library Cylinder Audio Archive
I strongly recommend exploring some of the audio available here: https://cylinders.library.ucsb.edu/index.php. These include a series of curated playlists, such as Patrick Feaster's selection of proto-radio drama in 'Audio Theatre', and collector John Levin's 'Recorded Incunabula 1891–1898'. This includes material we've touched on before, such as 'Why Should I Keep From Whistling?' by John Yorke Atlee, as well some more

obscure releases. 'Just One Girl', recorded for the Kansas City Talking Machine in 1898, is a good example of some of the questionable quality if customers strayed too far from the dominant forces of the day – Edison and Columbia. Give it a listen – there's speed deviation in the recording, and the cylinder finishes before the song does.

United States Phonograph Company
By the spring of 1896, Edison Spring Motor Phonographs were offered for sale with motors designed by Frank Capps of the United States Phonograph Company. Edison purchased the company and continued using their motors for many years.

US Everlasting Cylinders
The US Phonograph Company of Cleveland, Ohio, produced both two- and four-minute cylinders under its own 'Everlasting' label, in three series: popular, foreign language and grand opera. Like Blue Amberols, these were made of celluloid.

V-Disc
Record label that was formed in 1943 to provide records for US military personnel.

Velvet Face
Made by J. E. Hough's Edison Bell Company, these were marketed as a high-class, low-noise product, but it is doubtful whether Edison Bell used a different shellac mix for their production. These were replaced by the superior 'Electron' records.

Vitaphone
This was one of the very first disc records to be sold in Britain, c.1900. They were made by the American Talking Machine Company and were 7 inches in diameter and pressed in a red/brown material, with details printed in gold in the central area.

Victrola
The first 4ft high Victrola was made in 1906. Even though it was priced at $200, it was an immediate success – 200 orders were wired to dealers within days of its unveiling.

Virtual Gramophone: Canadian Historical Sound Recordings
A multimedia website devoted to the early days of Canadian recorded sound, providing an overview of the 78rpm era in Canada. The database contains 78rpm and cylinder recordings released in Canada from 1900 to 1950, as well as foreign recordings featuring Canadian artists and/or compositions. Each entry provides information about an original recording, such as its title and performer, relevant dates, and details about the label and disc. You can also explore the first Canadian record companies, including Berliner's Gram-o-phone Company of Canada and the Compo Company Limited.

Vitrolac
RCA launched coarse groove discs of 'Vitrolac' vinyl plastic running at 33⅓rpm in 1931, but the venture failed. Columbia's vinylite records appear in 1948, and the RCA Victor 7-inch 45rpm microgroove record the following year.

Vocalion
This British company started in 1920 and was initially called Aeolian-Vocalion. The label design was similar to the already-established American label and there were numerous catalogue series and owners over the next three decades. The company introduced its popular 8-inch Broadcast records in 1927, increasing to 9-inch in 1931.

Volta Lab
Set up by Alexander Graham Bell, this is where Massachusetts-born engineer Charles Tainter made the first lateral-cut or

zigzag record in the summer of 1881. The National Museum of American History has made all the digitised recordings available – just search by 'Volta Lab recordings'. For a more detailed guide to each recording, try the First Sounds collective: firstsounds.org/sounds/volta/.

Wangemann, Theo
You can listen to recordings made by Thomas Edison's West Orange lab assistant Theo Wangemann. Theo was sent to Europe in 1889 to display the phonograph at the Paris Expo. The trip was extended and he made a number of historic recordings, including the voice of Otto von Bismarck, accessed via: www.nps.gov/edis/learn/photosmultimedia/theo-wangemann-1889-1890-european-recordings.htm.

Waikiki Hawaiian Orchestra
Very late Blue Amberols were dubbed from electrically recorded discs and sound like the more familiar electrical recordings of the 1920s. You can hear an example of an electrically recorded Blue Amberol via https://cylinders.library.ucsb.edu – specifically 'Honolulu Moon' by the Waikiki Hawaiian Orchestra.

Woolworths
Many big chains and department stores did, at one time or another, produce their own labels. Woolworths had lots – the 9-inch 'Crown', the six-penny 'Little Marvel', and the 8-inch 'Eclipse' records. More well known were the 4s 6d 'Embassy' records, recorded by Levy's Oriole Company at their studios in 73 Bond Street, and pressed at the Oriole factory in Aylesbury, Bucks. Embassy records were high-quality cover recordings of popular chart hits. Some famously outsold the originals: the Maureen Evans version of 'Stupid Cupid' on Embassy outsold the Connie Francis original on MGM. Sales were generally very

high, which may be why Embassy record sales became excluded from the published chart 'Top 20'. Embassy records continued until 1965, when Oriole was taken over by CBS.

World
This was inventor Noel Pemberton-Billing's attempt to improve recording quality and extend the playing time of records. Similar to the aforementioned 'Longanote' discs, he tried to make the velocity of needle on groove constant by slowing it down at the start of the disc but gradually speeding up as it came towards the centre. This actually increased the playing time of the record – starting slowly, reaching 80rpm by the end. However, the discs were not successful – there were no recording stars on his roster, the records were expensive and you needed a special gadget to play them.

Wurlitzer
Wurlitzer's success was built around the Simplex record changer – the first could only manage a modest 10 78rpm discs, but this soon expanded to 24. Around the same time the model A 'Rock-Ola' was introduced, which could manage 12 discs, soon upped to 20 in the IMP-20 models.

Zonophone
Originally an American company, the Zonophone name was bought by the Gramophone Company in 1903 and it started issuing single-sided discs of 5½, 7, 10 and 12 inches. In 1913, the name was applied to the double-sided records that had been called 'Twin', and both names appeared on the label for some years. In 1932, when Columbia and the Gramophone Company were merged to form EMI, Zonophone was once again paired with an existing label and Regal-Zonophone was born. This survived until 1948. The Zonophone name was also revived in the 1960s.

Bibliography

Barnum, P.T. 1871. *Struggles and Triumphs: or, Forty Years' Recollections of P. T. Barnum*. American News Company, New York, USA.

Bell, A.G. Undated. Making a talking machine. Library of Congress. www.loc.gov/resource/magbell.37600301.

Bell, A.G. 1880. On the production and reproduction of sound by light. *American Journal of Science*, October 1880. New York, USA.

Bolig, J.R. 2008. *The Victor Black Label Discography*. Mainspring Press, Denver, USA.

Bragg, W. 1921. *The World of Sound*. Bell, London, UK.

Brooks, T. 2004. *Lost Sounds: Blacks and the Birth of the Recording Industry*. University of Illinois Press, Chicago, USA.

Burrows, T. 2017. *The Art of Sound: A Visual History for Audiophiles*. Thames & Hudson, London, UK.

Carson, B.H., et al. 1949. A Record Changer and Record of Complementary Design. *RCA Review*, June 1949. Radio Corporation of America, New Jersey, USA.

Chew, V.K. 1981. *Talking Machines*. HMSO, London, UK.

Coleman, M. 2005. *Playback: From the Victrola to MP3, 100 Years of Music, Machines, and Money*. Da Capo Press, Boston, USA.

Cramer, A & Koenigsberg, A. 1992. The World's Oldest Recording: Frank Lambert's Amazing Time Machine. A*ntique Phonograph Monthly* 10(3), New York, USA.

Cros, C. 1877. Letter deposited at the French Academy of Sciences. www.nps.gov/edis/learn/historyculture/origins-of-sound-recording-charles-cros.htm.

DeGraaf, L. 2013. *Edison and the Rise of Innovation*. Sterling, New York, USA.

Edison, T. 1888. The Perfected Phonograph. *The North American Review*, June 1888. New York, USA.

Feaster, P. 2007. *The Following Record: Making Sense of Phonographic Performance, 1877–1908*. Indiana University, Bloomington, USA.

Feaster, P. 2009. The Phonautographic Manuscripts of Édouard-Léon Scott de Martinville. FirstSounds.org.

Feaster, P. 2010. Édouard-Léon Scott de Martinville: An Annotated Discography. *Association for Recorded Sound Collections Journal* 41(1). Annapolis, USA.

Feaster, P. 2012. Phonographic Treasures of the Smithsonian. *The Antique Phonograph*. 30 (1, 2 & 4). Washington D.C., USA.

Feaster, P. 2015. A Discography of Volta Laboratory Recordings at the National Museum of American History. Smithsonian National Museum of American History, Washington D.C., USA. octaverter.com/pdf/volta-discography.pdf.

Feaster, P. 2016. The First Phonautograph in America. griffonagedotcom.wordpress.com/2016/05/25/the-first-phonautograph-in-america-1859/.

Feaster, P. 2017. Speed-Correcting Phonautograms Without Pilot Tones. griffonagedotcom.wordpress.com/2017/12/10/speed-correcting-phonautograms-without-pilot-tones/.

Feaster, P. 2017. The Phonograph as Toastmaster. griffonagedotcom.wordpress.com/2017/04/05/the-phonograph-as-toastmaster-october-5-1888/.

Feaster, P. 2017. Daguerreotyping the Voice: Léon Scott's Phonautographic Aspirations. griffonagedotcom.wordpress.com/2017/04/23/daguerreotyping-the-voice-leon-scotts-phonautographic-aspirations/.

Feaster, P. 2020. What Are The Wild Waves Saying?: Logograph Records of Spoken English, 1873–1878. griffonagedotcom.wordpress.com/2020/09/15/what-are-the-wild-waves-saying-logograph-records-of-spoken-english-1873-1878/.

Gelatt, R. 1977. *The Fabulous Phonograph 1877–1977*. Cassell, London, UK.

Giovannoni, D. 2017. Who Invented Sound Recording? Thomas Edison National Historical Park, New Jersey, USA. www.nps.gov/edis/learn/historyculture/origins-of-sound-recording-the-inventors.htm.

Goldmark, P.C. 1973. *Maverick Inventor: My Turbulent Years at CBS*. E.P. Dutton, New York, USA.

Hafner, K. 1999. In Love With Technology, as Long as It's Dusty. *New York Times*, 25 March, USA.

Hall, S. 2013. Indigenous American Cylinder Recordings and the American Folklife Center. blogs.loc.gov/folklife/2013/11/indi genous-american-cylinder-recordings-and-the-american-folklife-center/.

Hanley, R. 1977. Centenary of Edison Phonograph Marked. *New York Times*, 13 August, USA.

Harvith, J. & Harvith S.E. 1987. *Edison, Musicians, and the Phonograph*. Greenwood Press, Westport, USA.

Helmholtz, H.V. 1895. *On the Sensations of Tone as a Physiological Basis for the Theory of Music*. Longmans, Green & Co., London, UK.

Henry, E. 1888. The Graphophone. Paper read at the British Association for the Advancement of Science Bath Meeting. www.aes-media.org/historical/html/recording.technology.history/ar312.html.

Hepworth, D. 2019. *A Fabulous Creation: How the LP Saved Our Lives*. Bantam Press, London, UK.

Howard, J.T. & Bellows, G.T. 1957. *A Short History of Music in America*. Thomas Y. Crowell, New York, USA.

Huffman, L. Undated. First Electric Recordings of Leopold Stokowski and the Philadelphia Orchestra. www.stokowski.org.

Huffman, L. Undated. Victor Program Transcription Records. www.stokowski.org.

Kilgarriff, M. Undated. The Voice of Henry Irving. www.theirvingsociety.org.uk.

Marmorstein, G. 2007. *The Label: The Story of Columbia Records*. Thunder's Mouth Press, New York, USA.

Maslon, L. 2018. *Broadway to Main Street: How Show Music Enchanted America*. Oxford University Press, Oxford, UK.

McClure, J.B. 1879. *Edison and his Inventions: including the many incidents, anecdotes and interesting particulars connected with the life of the great inventor*. Rhodes & McClure, Chicago, USA.

Millard, A. 1995. *America on Record: A History of Recorded Sound*. Cambridge University Press, Cambridge, UK.

Milner, G. 2009. *Perfecting Sound Forever: An Aural History of Recorded Music*. Faber, New York, USA.

Mitchell, O. 1924. *The Talking Machine Industry*. Pitman, London, UK.

Moore, J.N. 1999. *Sound Revolutions: A Biography of Fred Gaisberg, Founding Father of Commercial Sound Recording*. Sanctuary Publishing, London, UK.

Morton, D. 2000. *Off the Record: The Technology and Culture of Sound Recording in America*. Rutgers University Press, Piscataway, USA.

Newnham, G.L. 2019. Charles Pathé and the Phonograph. www.pathefilm.uk

Newville, L.J. 1959. Development of the Phonograph at Alexander Graham Bell's Volta Laboratory. *United States National Museum Bulletin*, 218. United States National Museum and the Museum of History and Technology, Washington D.C., USA.

Osborne, R. 2012. *Vinyl: A History of the Analogue Record*. Routledge, London, UK.

Perks, R. & Prentice, W. 2005. Recording Florence Nightingale's voice. *Playback* 33. British Library Sound Archive, London, UK.

Read, O. & Welch, W.L. 1976. *From Tin Foil to Stereo: Evolution of the Phonograph*. Howard W. Sams, Indianapolis, USA.

Rich, J. 2017. Acoustics on display: collecting and curating sound at the Science Museum. *Sound and Vision*, Spring 2017. Science Museum, London, UK.

Rodríguez, E.M. 2015. *Music and Exile in Francoist Spain*. Ashgate, London, UK.

Rust, B. 1975. *Gramophone Records of the First World War: An HMV catalogue, 1914–1918*. David & Charles, Newton Abbot, UK.

Rust, B. 1977. The Development of the Record Industry in USA. *Gramophone*, April 1977. London, UK.

Rust, B. & Brooks, T. 1999. *The Columbia Master Book Discography*. Greenwood Press, London, UK.

Sagan, C. et al. 1978. *Murmurs of Earth: The Voyager Interstellar Record*. Random House, New York, USA.

Schoenherr, S.E. 1999. Charles Sumner Tainter and the Graphophone. Recording Technology History. www.aes-media.org/historical/html/recording.technology.history/graphophone.html.

Scott, J. 2019. *The Vinyl Frontier: The Story of the Voyager Golden Record*. Bloomsbury Sigma, London, UK.

Scott de Martinville, E.L. 1857. *Principes de Phonautographie*. Facsimile edition by David Giovannoni. firstsounds.org.

Seymour, H. 1918. *The Reproduction of Sound: Being a Description of the Mechanical Appliances and Technical Processes Employed in the Art*. W.B. Tattersall, London, UK.

Sooy, H.O. 1925. *Harry O. Sooy Memoir*. Photostatic copy of typed memoirs. Hagley Museum and Library, Wilmington, USA. digital.hagley.org/sooybros.

Sousa, J.P. 1906. The Menace of Mechanical Music. *Appleton's Magazine* 8. Appleton & Company, New York, USA.

Sterne, J. 2003. *The Audible Past: Cultural Origins of Sound Reproduction*. Duke University Press, Durham, USA.
Sutton, A. 2010. *A Phonograph in Every Home: The Evolution of the American Recording Industry*. Mainspring Press, Denver, USA.
Thompson, C. 2016. How the Phonograph Changed Music Forever. *Smithsonian Magazine*, April. Washington D.C., USA. www.smithsonianmag.com/arts-culture/phonograph-changed-music-forever-180957677/.
Thompson, E. 1995. Machines, Music, and the Quest for Fidelity: Marketing the Edison Phonograph in America, 1877–1925. *The Musical Quarterly*, vol. 79, no 1. Oxford University Press, Oxford, UK.
Vander Lugt, M. 2017. The Recordings of the Columbia Phonograph Company, 1889–1896. Association for Recorded Sound Collections. arsc-audio.org/blog/2017/05/23/columbia89-96/.
Wile, R.R. 1982. The Edison Invention of the Phonograph. *Association for Recorded Sound Collections Journal* 14(2). Annapolis, USA.
Wile, R.R. 1990. Etching the Human Voice: The Berliner Invention of the Gramophone. *Association for Recorded Sound Collections Journal* 21(1). Annapolis, USA.
Wile, R.R. 1991. Edison and Growing Hostilities. *Association for Recorded Sound Collections Journal* 22(1). Annapolis, USA.
Wile, R.R. 1993. The Launching of the Gramophone in America, 1890–1896. *Association for Recorded Sound Collections Journal* 14(2). Annapolis, USA.
Wile, R.R. 1996. The Gramophone Becomes a Success in America, 1896–1898. *Association for Recorded Sound Collections Journal* 27(2). Annapolis, USA.
Wilentz, S. 2012. *360 Sound: The Columbia Records Story*. Chronicle Books, San Francisco, USA.

Other sources and further reading

78rpm Club. Survey of 78-producing record companies and labels from across the world. 78rpm.club.
British 78rpm record labels. www.mgthomas.co.uk/Records/labelindex-A.htm.
Charles Sumner Tainter Papers. National Museum of American History, Archives Center, Washington, D.C., USA. sova.si.edu/record/NMAH.AC.0124.
Early Record Catalogues. Collection of catalogues for the British market from 1898 to 1926, including ACO, Columbia, Edison,

Edison-Bell, Gramophone, His Master's Voice, Homochord, Monarch, Odeon, Pathe, Regal, Vocalion and Zonophone. British Library, London, UK. sounds.bl.uk/Sound-recording-history/Early-record-catalogues.

Echoes of History. Association for Recorded Sound Collections podcast. www.arsc-audio.org/echoes-of-history.html.

Edison Phonograph Monthly. Trade magazine published from 1903. National Phonograph Company, New York, USA. archive.org/details/edisonphonograph01moor.

Emile Berliner and the Birth of the Recording Industry. Recordings and manuscripts held by the Recorded Sound Section of the Motion Picture, Broadcasting and Recorded Sound Division of the Library of Congress, Washington D.C., USA. www.loc.gov/collections/emile-berliner/.

First Sounds. Listing of publications available through the First Sounds collective. www.firstsounds.org/publications.

Hillandale News. Journal of the City of London Phonograph and Gramophone Society. https://archive.org/details/thehillandalenews.

Talking Machine World. Trade magazine published from 1905. Edward Lyman Bill, New York, USA. mediahistoryproject.org/reader.php?id=talkingmachinewo05bill.

The Edison Disc. City of London Phonograph and Gramophone Society article. www.clpgs.org.uk/uploads/4/9/3/8/49389291/_the_edison_disk.pdf.

The First Book of Phonograph Records: 1889–92. Digitised logbook of the first musical recording program at the Edison Laboratory. https://archive.org/details/FirstBookOfPhonographRecords/page/n5/mode/2up.

The Phonogram. Trade title edited by Virginia McRae between 1891 and 1893. North American Phonograph Company, New York, USA. archive.org/details/Phonogram1_1.

The Phonoscope. Launched and edited by Russell Hunting from 1896. Phonoscope Publishing, New York, USA. archive.org/details/phonoscope13hunt.

Thomas Edison Papers Digital Edition. Rutgers, The State University of New Jersey, New Jersey, USA. edisondigital.rutgers.edu.

Scientific American. First reports of the invention of the phonograph in November and December 1877 issues of *Scientific American*: www.aes-media.org/historical/html/recording.technology.history/afteredison.html.

Acknowledgements

In the wake of my previous book, *The Vinyl Frontier*, a research chemist from the Library of Congress got in touch. We exchanged high-level geekery on a question of the Golden Record's final running order, which, in short, led to a hugely valuable and enjoyable research trip to America. I visited Washington, D.C., amid lockdowns and shutdowns. International travel was coming to an end, public buildings were closing, and shops would soon follow. Washington itself had a strange atmosphere. There were still plenty of people about, but there was an after-party mood in the air. This was near start of the Covid 19 pandemic, where panic rubbed shoulders with bemused excitement.

The chemist in question is a delightful chap called Andrew Davis. He works in the Library of Congress Preservation Research and Testing Division. This is a department that investigates what time does to stuff, and seeks ways to stop it happening. The Library of Congress looks after priceless books, manuscripts, instruments, gizmos, inventions ... there's a lot. So a chemist in the Preservation Research and Testing Division might be working on any number of projects, or within numerous specialisations. One might study the long-term deterioration of glass, for example, while another might study the impact of verdigris pigment, a pigment used in antiquity whose acidity is such that it contributes to the deterioration of paper.

On the day of my visit I was slightly nervous. I was greeted by the department chief, Dr Fenella France, an international specialist on environmental deterioration of cultural objects, who has worked on projects relating to preserving World Trade

Center artefacts and material held by the Ellis Island Immigration Museum. She is perhaps best known for using non-invasive spectral imaging to reveal hidden text and information, through which she discovered that Thomas Jefferson changed the word 'subjects' to 'citizens' in a rough draft of the Declaration of Independence. As I say, I was slightly nervous.

We did that awkward thing people did at the start of the Pandemic, namely rub elbows. In reality we didn't even do that, we just made a show of rubbing elbows, discussing how even then advice was changing, and people were being discouraged from any contact at all. Taking part in the elbow-off was Fenella, Andrew and another gentleman I had never met before, the technical lead of the IRENE system, Peter Alyea.

It was Peter I had come to see. He was at the helm of what had started several years before, when physicists at Lawrence Berkeley National Laboratory began experimenting with high-resolution microscopes as a method of scanning grooves in order to extract sound. Just as Fenella had used non-invasive spectral imaging to dig into hidden messages within fragile documents, so this technique could be used to unlock sound from fragile and unplayable grooved media, without damaging or even touching the surface. The project became a collaboration with the Library of Congress, and slowly developed methods to preserve the Library's sound collections.

The place where Peter works is so cool. I mean, the Library itself is amazing. It's a maze of doors, staircases, lifts and corridors that might put you in mind of a Romero movie from the 1970s. Peter's domain is a kind of open laboratory. We stood together, Peter generously answering questions he must have been asked countless times before, while giving off the pleasingly professorial air of someone who really knows their stuff. In the flesh I saw some of the objects I'd been reading about – not only IRENE's precision turntables and spindles, cameras and imaging equipment, but archaic discs, audio curios and cylinders, including numerous artefacts from Alexander Graham Bell's Volta Laboratory. Seeing these in the flesh, close enough to touch, and meeting people dedicated to using science to preserve and unlock audio history, was deeply inspiring.

I'd like to thank Jim Martin of Bloomsbury for the extensions and Lucy Beevor for the corrections. Thanks to Michael Thomas, whose exhaustive survey of 78rpm record labels in Britain (at www.mgthomas.co.uk/Records/RecordIndex.htm) formed the basis of many entries in the 'Miscellany of the Groove'. I'd also like to thank my paternal grandfather, Tim. Thanks to his slavish readership of the magazines *Wireless World*, *Hi-fi News* and *Gramophone*, I was able to peruse almost complete runs of all three titles at my leisure. These were invaluable for providing a contemporary, on-the-ground picture of the development of grooved media in the twentieth century, the popular obsession with high fidelity, and a conveyor belt of weird gizmos. Tim died in 1991, around the time I was discovering the work of Ned's Atomic Dustbin. He was an opera nut and audiophile. Sadly the gulf between the Neds and Wagner was too vast, meaning that during his lifetime we never bonded on the topic of music, which feels like an opportunity missed. But having his collection of magazines easily to hand meant I could read them during gaps in childcare.

Thanks to Genevieve (11), Florence (13) and Rupert (17) for growing up so quickly and providing gaps in childcare. Thanks to my partner Vanessa, for everything, but in particular for fielding my requests for reassurance. Thanks to Andrew, Fenella and Peter, and everyone else at the Library of Congress Preservation Research and Testing Division for being so generous with time. Thanks to the Eccles Centre for American Studies at the British Library. I'd also like to pay tribute to the work of Patrick Feaster. If you've enjoyed this book, I recommend searching out Patrick's work. He's written in granular detail on almost every conceivable topic relating to the early capture of sound waves. The fact that I'm writing this isn't to suggest he endorses this book by the way. Lord no, I am simply doffing my cap.

Index

Abbott, Daisy 158
Académie des Sciences 19, 27, 51, 52
'accidental stereo' recordings 239
acetate/lacquer discs 193–195, 197
acoustic era of recording 15–17, 160–176, 188
acoustic telegraphy 48, 57–58
acoustics 21–24
Adams, James 57, 63
Ader, Clément 238
AEG 236
Amberol cylinders 149–150, 155, 156–157, 230
Amberola 155
American Council of Learned Societies 125
American Graphophone Company 89–90, 94, 110, 111, 135
Ampex Electrical and Manufacturing Company 236–237
Argus 74
Armstrong, Louis 187
Asahi Sonorama 251
Associated Glee Clubs of America 184
Atlee, John York 105, 106, 107
Audio Engineering Society 248
Audio Fidelity Records 241–242
Audion 178
automatons 34–40
AVI Records 251
Aylsworth, Jonas W. 152–153

Baby Grand Gramophone 132
Baby Grand Graphophone 100
Bachman, Bill 216
Bacigalupi, Peter 144–145
Bailey, Mildred 187
Bancker, Charles N. 29
Band of the Grenadier Guards 202
Barlow, William Henry 32–33
Barnett Samuel 140, 201
Barnum, Phineas Taylor 34, 38, 39, 68
Barraud, Francis 171, 172
BASF 236
Batchelor, Charles 47, 48, 56–57, 58–59, 65, 74
BBC 193
Beatles 251
Beecham, Sir Thomas 240
Bell, Alexander Graham 17, 19, 39–40, 44, 58, 78–88
Bell, Alexander Melville 39, 77, 82, 84
Bell, Chichester 80, 86–87

Bell, Melville 39–40
Bell Laboratories 180–181, 196, 200–201, 239, 240
Bell records 175
Bell Telephone Company 127
Beneshan, Conrad 254–255
Berliner, Emile 126–136, 162, 238
Berliner, Joseph 164
Berliner Gramophone Company 133
Bettini, Gianni 158–160, 162
bit depth 263
bit rate 263
Blakely, David 102
Bliss, George 65
Blue Amberols 150, 155, 156–157, 230
Blumlein, Alan 239–240
Borwick, John 201
Boston Cadet Band 109
Boston Symphony 175
Boswell Sisters 187, 202
Brandenburg, Karlheinz 262
brass discs 86
Broadley, Alexander Meyrick 115
Brown, Milton 187
Brunswick Panatrope 200
Brunswick Records 186, 202
Bülow, Hans von 92

Calahan, Edward 47
Calloway, Cab 186, 202
Calvé, Emma 167
Capitol 99, 217, 223, 228
Capp's Seventh Regiment Band of New York 103
carbon button transmitter 79
carbon microphones 127, 179–180, 238
Cardiff Parks Committee 146
Carl Lindström Company 175–176
Carter, Jimmy 13–14, 20
Carter Family 186
cartridges, 8-track 243, 244
Caruso, Enrico 165–167, 173, 174
Case, Anna 154
'Casey' Records 109–110
cassettes 243–244
cast albums 227–228, 229
CDs 261–262, 263
celluloid 129, 130, 150, 156, 190
cellulose nitrate 193–194
Chaliapin, Feodor 167
Chess Records 237
Chevalier, Albert 166
Chew, Victor Kenneth 140

INDEX

Chicago World's Fair 108
Chladni, Ernst 22–24
Chladni figures 23
Christian, Charlie 187–188
Chronophone 195
Cinch records 151
Cinemacrophonograph 195
City of London Phonograph and Gramophone Society 144
Clarion cylinders 149
Clark, Alfred 164
Clearaudio 247
Climax records 134
clocks, speaking 59, 69, 70
Coborn, Charles 119–120
coin-in-the-slot phonographs 95–96, 99, 107, 108, 134, 137–139, 143
Collins, Arthur 104, 154–155
colour photography 51–52
Columbia 94, 99, 102–103, 105, 106–107, 108–109, 110–111, 134, 136, 141, 148–149, 151, 155, 175, 176, 179, 181, 182, 184–185, 209–217, 222, 224–225, 227–229, 231–234, 240
Como, Perry 228
compact cassettes 243–244
Concert cylinders 122, 143, 148–149
condenser microphones 180–181, 212
Congressional Record, The 90
Cook, Emory 240–241
Cooke, George Alfred 37
Coquet phonograph 122
Cornell, Earl 31
Cortot, Alfred 162, 182
Cosmocord 247
Cosmos 29
cowboy songs 124–125
Crawford, Jesse 180
Cros, Charles 18–19, 49–54, 62, 79, 81
Crosby, Bing 186–187, 202, 236–237
Crossley, Ada 174
Crystal Palace, London 116
Crystalate 202–203
Culshaw, John 203–204
currency, British 121
cylinder phonographs 60–78, 82–83, 87–88, 90–97, 98–111, 155
cylinders 14, 15–17, 147–150, 155–157

Darby, William Sinkler 131, 164
de Callias, Nina 50
de Forest, Lee 178, 195–196
Debussy, Claude 50
Decca 140, 186–187, 201–204, 217, 222, 227, 228, 229, 243
Decca Dulcephone 201
Der Ring des Nibelungen 203–204
Deram Label 243
Dickson, William 91, 195
digital era of recording 160
disc phonographs 72, 75, 81–83, 128–136

disco discs 250–251
discs 14, 15–16, 147–148, 150–155
 7-inch 132, 162, 214, 219–225
 16rpm speed 20, 250, 255, 258
 33rpm speed 194, 196–197, 209, 220, 222
 45rpm speed 220–225, 250
 78rpm speed 131–132, 194
 brass 86
 CDs 261–262, 263
 disco 250–251
 EPs 250
 flexi 251
 glass 85–86, 128
 lacquer/acetate 193–195, 197
 LPs 198–199, 205–217, 219, 222, 226–229, 233, 234
 masters 190–191, 194–195, 197
 pressing process 190–192, 194–195, 197, 214
 rubber 129, 130, 132–133
 shellac 14, 20, 133, 189–192, 207, 214
 sleeves 230–234
 vinyl 14, 20, 210–217
 zinc 129, 130, 131
Doegen, Wilhelm 124
dolls, talking 59, 69, 130
Donovan, Dan 107
Downbeat Magazine 188
Drake, Frank 255–258, 260
Druyan, Ann 259–260
Ducos du Hauron, Louis 52
Duhamel, Jean-Marie Constant 24
Dunn, Bob 187
Duranoid Company 133
Durham, Eddie 187
dust jackets 229–230
Dylan, Bob 225
Dynagroove 249

Eames, Emma 174
Easton, Edward 98–99, 111
Edison, Charles 200, 206
Edison, Marion 61–62
Edison, Nancy 45
Edison, Samuel 45
Edison, Thomas 8–9, 14–15, 17–18, 41, 42, 44–48, 53, 55–65, 66–71, 75–76, 77–79, 81, 88, 89–97, 111, 114, 115, 153, 195, 205–206, 230
Edison Amberols 149–150, 155, 156–157, 230
Edison Bell Phonograph Company 149, 175
Edison Concert Records 122, 143, 148–149
Edison Diamond Discs 153, 156
Edison Gem 122
Edison Gold Moulded Records 143, 148, 152
Edison Grand Concert Band 123

Edison Long Play   206–207
Edison Perfected Phonograph   91, 114, 118
Edison Phonograph Monthly   123, 142, 152, 154
Edison Speaking Phonograph Company   69, 75, 78, 88, 93, 148–150, 152–157, 199–200
Edison Standards   105, 150
Edmunds, Henry   74
electric era of recording   160, 177–188
Electrical World   77–78
Emerson, Victor   105
EMI   136, 240, 243
English Mechanic   74
EPs (extended-play records)   250
Euler, Leonhard   36
Euphonia   37–39
Everlasting cylinders   155–156
Express   141

F. Ruppel & Co   120
Faber, Joseph   37–39, 68
Fantasound   240
Favorite records   151
Fay, Harry   163, 202
Feaster, Patrick   29–30, 114
Fenby, F. B.   54
Ferrier, Kathleen   202
Ferris, Timothy   252, 259, 260
Fessenden, Reginald   178
Fewkes, Jesse Walter   123
Filmophone records   190
films, with sound   195–197
Fine, C. Robert   242–243
'First Book of Phonograph Records'   96
First Sounds collective   31–32
Flaunty, William   140
Flaunty's Phono Stores   140
Fletcher, Alice Cunningham   123
Fletcher, Harvey   239
flexi discs   251
Ford Wayne Sentinel   122
Formby, George   202
Franklin, Benjamin   22
Franklin Institute, Philadelphia   129
Franz Josef, Emperor of Austria   236
Frey, Sidney   241–242
Fritz Puppel   140–141
Full Frequency Range Recordings   203

Gaisberg, Fred   105–109, 130–131, 132, 162, 163–167
Gaisberg, William   165, 167
Galilei, Galileo   13, 21–22
Garfield, Andrew   81
Garland, Judy   240
Garrard   200
Gaskin, George   108, 132
Gaumont, Léon   195

Gelatt, Roland   93, 94, 156
Gennett Record Company   176
Germain, Sophie   23–24
Gershwin, George   170
Gilliam, Art   182
Gilmore, Patrick   96
Giovannoni, David   31–32
Glass, Louis   95
glass armonica   22
glass discs   85–86, 128
Globe Record Company   134
Gloetzner, Raymond   131
Golden, Billy   104, 108, 130
Goldmark, Peter   13–14, 20, 21, 199, 205, 208–217, 221
Goodman, Benny   185, 187–188
Goossens, Marie   242
Gouraud, Colonel   112–118
Graham, George   132
gramophone   126–136, 162, 201
Gramophone   201, 204
Gramophone Company Limited   136, 140, 150, 151, 162–169, 171–175, 182
Grand Opera Amberols   150
Grand Trunk Herald   46
Grand Wizzard Theodore   253
graphophone   77–88, 89–90, 92–94, 98–100, 111, 128, 134, 136
Graphophone Grand   122
Gray, Elisha   44, 48
Green, Charlotte   31
Grey Gull Records   207
Guest, Lionel   179–180
Guiteau, Charles J.   81

Haber, Carl   31
Haddon, Alfred Cort   123–124
Haddy, Arthur   202–203
Hammerstein, Oscar   226–227
Hammond, John   188
Hampton, Lionel   242
Handy, W. C.   176
Harper's Weekly   68
Harris, Augustus Glossop   114–115
Harrison, Henry C.   180
Harrods   139
Haussmann, Georges-Eugène   26
Heine, William K.   248
Heineman, Otto   176
Henderson, Roy   202
Henry, Joseph   38
Hertz, Heinrich   178
hill-and-dale recording   81, 83, 128, 151–152, 156
'His Master's Voice' painting   171–173
HMV   153, 172, 173, 240
Hobbies magazine   154
Hofmann, Josef   92
Hollingshead, John   39
Homophone records   151

INDEX

Hooke, Robert 22
Horowitz, Vladimir 208
Hough, James E. 149, 151
Hubbard, Gardiner Greene 78, 80
Hughes, David 238
Hunter, Jim 216
Hunting, Russell 98, 109–110, 165, 182

Imhof's 139
indecent recordings 110
Indestructible Cylinders 155
indigenous songs and stories 123–124
International Exposition of Electricity, Paris 238
IRENE project 84–86
Irving, Henry 116–117

Jefferson, Thomas 102
Johnson, Edward H. 60–61, 64, 67, 73, 76
Johnson, Eldridge R. 131–132, 134, 135, 165, 171–172, 174
Johnson, George W. 105, 109
Jolson, Al 186, 197, 202
Jones, Joseph 135
Jones, William C. 133
Journet, Marcel 174
Jumbo records 151

Kämmer & Reinhardt 130
Kapp, Jack 143, 186–187, 227
Kell, Reginald 242
Keller, Arthur C. 200–201
Kempelen, Wolfgang von 35–36, 37
Kesten, Paul 211
Kinetophone 195
Kipling, Rudyard 165
Kittredge, George Lyman 124
Koenig, Rudolph 30–31
Koenigsberg, Allen 133
Koltzow, A. 120
Koussevitzky, Serge 174, 175
Kratzenstein, Christian Gottlieb 36
Kruesi, John 48, 56–57, 61, 63, 64

La Flesche, Francis 123
lacquer/acetate discs 193–195, 197
Lake County Star 51
lampblack 128
Lanfried, Martin 117
language courses 99
laser turntables 248–249
lateral-cut recording 24, 81, 128, 130–131, 156, 176, 218–219
Lawrence Berkeley National Laboratory 31, 84
Le Petit Journal 43
Ledbetter, Huddie 125
Lee, Peggy 228
Lennon, John 252
Leno, Dan 164

Lenoir, Abbé 62
Lewis, Edward 201–202
Library of Congress 18, 83, 84, 123
Lieberson, Goddard 227–229
Liebler, Vin 216
Lindsay, Bruce 139
Lippincott, Jesse 92–94
Logograph 32–33
Lomax, Alan 125
Lomax, John 124–125
Lombardo, Guy 185
LPs (long-playing records) 198–199, 205–217, 219, 222, 226–229, 233, 234
Lyons, Joseph 128

McClure, James Baird 66, 71, 172–173
McCrory, John 124
Macdonald, Thomas Hood 107
Mackenzie, James 46
Madame X discs 217, 219–225
magnetic cartridge 245–246
magnetic era of recording 160, 235–244
Manet, Édouard 50
Manzetti, Innocenzo 43
Marathon records 150, 207
Marconi, Guglielmo 74, 193
Marsh, Orlando 180
Marshall, Charles 103–104, 105
Marx, Groucho 226
Maskelyne, John Nevil 36–37
Mason, Thomas 73
master matrix 191
masters 190–191, 194–195, 197
Mauro, Philip 88
Maxfield, Joseph P. 180
Melba, Nellie 167, 193
Meller, Vlado 252, 259
Memphis Recording Studio 224
Menlo Park, New Jersey 48, 56–65, 66–69
Mercury Records 217, 242–243
Merriman, Horace 179–180
Metropolitan Opera House 184
Meucci, Antonio 43–44
Meyers, Johnny 109
Michailowa, Marie 174
microgrooves 20, 214, 219, 233
microphones 127, 179–181, 183, 212, 238, 239
Miller, Glenn 185
Miller, Walter H. 152–153, 206
Miller, William 34
Millophone records 150
Mills Brothers 186, 202
Milstein, Nathan 214, 233
Mitthauer, John 96
Moreschi, Alessandro 165
Morse, Samuel 42
Morton, Jelly Roll 180, 186
mother matrix 191
Motown Records 250–251

movies, with sound 195–197
Mozart, George 163
MP3 262–263
Mullin, John T. 236
Multiplex Graphophone Grand 122, 245
music boxes 69, 70
musicals 226–229
Musogram 150, 207

NASA 20–21, 83, 252–253, 255–260
National Association of Talking Machine Jobbers 153
Native American songs and stories 123
Neophone records 150, 152, 207
New England Phonograph Company 109
New York Gramophone Company 133
New York Phonograph Company 103–104, 105
New York Times 178
Nicole records 150
Nightingale, Florence 117–118
Nipper the dog 171–173, 174
noise reduction 249–250
North American Phonograph Company 93–94, 96–97, 99, 107
North American Review 70
Northants Talking Machine Society 153

Obert-Thorn, Mark 239
Ockenden, Frank 247
Odeon Records 151, 175–176, 231
Oink comic 251
Okeh label 176
Oklahoma! 226–227
Original Dixieland 'Jass' Band 176
O'Terrell, John 132
Owen, William Barry 136, 162, 171

paleophone 19, 49–54, 62
Paley, William 215, 217
Panama-Pacific Exposition, San Francisco 154
Parkins & Gotto 139
patents
 cylinder manufacture 150
 Edison's phonograph 64–65, 72, 75, 87, 88, 91, 92, 111, 118, 153
 Fenby's 'phonograph' 54
 gramophone 128, 133, 134–135
 graphophone 88, 89–90, 92, 111, 118, 134
 'indenting' versus 'incising' 88, 92
 microphones 127, 238
 paleophone 53–54
 telegraphs 47
 telephones 43, 44, 79
Pathé 122, 137–139, 149, 150, 151–152, 155
Pathé, Charles 138
Pathé, Emile 138

Patti, Adelina 167
Peckham, George 252
Peer, Ralph 186
Pepper's Ghost 37
Philadelphia Centennial Exhibition 58
Philadelphia Orchestra 175, 183, 198, 200–201, 239
Philadelphia Record, The 76
Philco Corporation 213
Philips, A. R. 160
Philips Electrical Ltd 243–244
Phillips, Sam 224, 237
Phoenix records 151
phonautograph/phonautograms 18, 25–33, 127
Phono-Cinéma-Théâtre 195
Phonofilm 196
Phonogram, The 95, 99, 104
phonographs 14–19
 coin-in-the-slot 95–96, 99, 107, 108, 134, 137–139, 143
 cylinder designs 60–78, 82–83, 87–88, 90–97, 98–111
 disc designs 72, 75, 81–83, 128–136
 early public interest in 66–74, 76
 Edison's 8–9, 14–15, 17–18, 19, 55–78, 90–97, 114, 135–136
 first uses of word 54, 62
 gramophone 126–136, 162, 201
 graphophone 77–88, 89–90, 92–94, 98–100, 111, 128, 134, 136
 invention of 17–18, 55–65
 IRENE project 84–86
 known as 'talking machines' 14, 35, 68–69
 leased 93–94, 99
 low-cost 15, 111, 121–122
 medical uses 71
 predicted uses 14–15, 70–71
 prices 15, 75, 111, 121–122, 143
 public events 145–146
 tin-foil recording 59–63, 69, 71–74, 80
 wax recording 59, 72, 80–83, 86, 87–88, 91–92, 149
Phonoscope, The 109–110, 160
Phonotrader and Recorder 141
photography, colour 51–52
'photography of sound' idea 26–28, 62
photophone transmitter 79
pick-up 245–246
Pini, Anthony 242
Pinto, Amelia 166
Polyphonwerk 202
Popular Science Review 33
Poulsen, Valdemar 236
Pozo, Antonio 100–101
Practical Applications of Electricity 32
Preece, William 74
Prescott, George 8
Presley, Elvis 224, 237

INDEX

pressing process 190–192, 194–195, 197, 214
prices
  phonographs 15, 75, 111, 121–122, 143
  records 142, 143, 151, 157
Program Transcription records 197–199, 209–210
Puck phonograph 15, 120–122

quadraphonic records 249
quadruplex telegraph circuit 47
Quinn, Daniel 109, 132

radio 157, 177–179, 192–193, 197, 207
Rainey, Ma 176
RCA 178, 197–199, 207, 209, 215, 217, 219–224, 228, 229, 249, 250
Read, Oliver 162
record shops 139–145
record sleeves 230–234
Recording Industry Association of America (RIAA) 262
recording methods
  hill-and-dale 81, 83, 128, 151–152, 156
  'indenting' versus 'incising' 88, 92
  lateral-cut 24, 81, 128, 130–131, 156, 176, 218–219
  on tin-foil 59–63, 69, 71–74, 80
  on wax 59, 72, 80–83, 86, 87–88, 91–92, 149
record-making process 190–192, 194–195, 197, 214
Red Seal records 173–175
Reiner, Fritz 221
Reis, Philipp 43
Reisman, Leo 185
'Rhapsody in Blue' 170, 176, 184
Richard, Cliff 244
Riddle, Frederick 242
Rodgers, Jimmie 186
Rodgers, Richard 226–227
Rodman, Ike 216
Rogers, Harry G. 86
Rose, Andrew 239
Rosenthal, Richard 99
Royal Institution 74
Royal Society of Victoria, Australia 74
rubber discs 129, 130, 132–133
Russian Academy of Sciences 36

Sagan, Carl 20, 252–253, 255–258, 259–260
Sagan, Linda Salzman 255–256
Saint-Saëns, Camille 183
Salkind, Saul 139
Salon du Phonographe 137–139
sample rate 262–263
Sanders, Joseph 131
Sanders, Zip 131
Sarnoff, David 178, 215, 217
Savory, Bill 216

Schulze-Berge, Franz 91
Schumaker, Edward 198
Science Museum, London 146
Scientific American 14, 60, 62, 64, 65, 67, 172
Scott, Robert Falcon 173
Scott de Martinville, Édouard-Léon 18, 25–28, 30–31, 32, 127
Scotti, Antonio 174
scratching 253
Seaman, Frank 133–134
Sembrich, Marcella 174
Seuss, Werner 131
Shaw, Alice 113
shellac discs 14, 20, 133, 189–192, 207, 214
Shepard, Bert 163
Shore, John 22
Sievey, Chris 251
Simpson, R. F. 189
Sinatra, Frank 228
sleeves 230–234
Smith, Bessie 176, 183
Smith, Mamie 176
Smithsonian Institution 82–83, 84, 127
Snepvangers, Rene 211, 212, 216
Sobinov, Leonid 165
Society of Telegraph Engineers 74
Sooy, Charlie 169–170
Sooy, Harry 169–170, 181, 182–183
Sooy, Raymond 169–170
Sound Wave 141
sound-on-film system 195–196, 197
soundtracks 229
Sousa, John Philip 15, 102–103, 106, 177
South Pacific 226, 228, 232
South Wales Daily 112–113
South Wales Echo 50
Southard, Paul 215
speech synthesizers 34–40, 68
speeds
  16rpm 20, 250, 255, 258
  33rpm 194, 196–197, 209, 220, 222
  45rpm 220–225, 250
  78rpm 131–132, 194
Spencer, Len 108
Spiller, Henry 140
Spillers Records 140
spring motors 120, 131–132
stampers 191–192, 214
stamps, talking 252
Stanton, Frank 213
Steinweiss, Alex 231–234
stereo 238–243
Stereophile 249
Sterling, Louis 182
Sterling cylinders 149
Stevens, Risë 221
Stevens Institute of Technology, Philadelphia 29, 30
Stewart, Cal 105

Steytler, Charles 117
Stilwell, Mary 47, 56
stock tickers 47, 58
Stoker, Bram 112
Stokowski, Leopold 174, 175, 183, 198, 200–201, 239
Stollwerck chocolate records 251–252
streaming 262–263
Strindberg, August 50
Sullivan, Arthur 114, 115, 165
Sun Records 224, 237
swear word, first recorded 85–86

Tainter, Charles Sumner 79–88, 90, 93–94, 107–108
talking dolls 59, 69, 130
Talking Machine News 141, 161, 167–169
talking machines 14, 34–40, 68–69, 136
talking pictures 195–197
talking stamps 252
Tamagno, Francesco 167
tape recording 235–237, 243–244, 261
telegraphs 41–43, 46–48, 57–58
telephones 17, 39, 41, 43–44, 58, 64, 70, 78, 79
Tenniel, John 121
Tennyson, Lord 117
Terry, Ellen 117
Théâtrophone 238–239
Thomas Edison National Historical Park 29
Thompson, Clive 104
Tigerstedt, Eric 195
Times, The 74
tin-foil recording 59–63, 69, 71–74, 80
Today (radio programme) 31
Todd, Burt 252
Tone Tests 153–155
tonearms 246–248
Toscanini, Arturo 174–175, 208
tracking force 246
Transit of Venus 50–51, 79
tuning fork 22, 24, 31
The Turk 35
Turner, Ike 224
typesetting 26

United States Gramophone Company 131
United States Marine Band 102–103, 109
United States Phonograph Company 155–156

Vail, Alfred 42
Vallée, Rudy 185–186
Variety 228–229
Velvet Face records 175
Venus, Transit of 50–51, 79
verrillon 22
vertical-cut recording *see* hill-and-dale recording
vibrograph 24

vibroscope 24
Victor Military Band 176
Victor Talking Machine Company 135, 136, 153, 155, 167, 169–170, 172, 179, 181, 182–183, 184–186, 198, 199–200
Victrola 136, 155
Victrolac records 197–199, 207
vinyl discs 14, 20, 210–217
Vinylite plastic 20, 210
Vitaphone 134, 196–197
Volta Graphophone Company 87, 89
Volta Laboratory 19, 80–88
vox humana 36
Voyager Golden Records 20–21, 83, 252–253, 255–260

Wagner, Richard 203–204
Wallerstein, Edward 199, 209, 212, 213, 215–216, 233
Walt Disney 240
Walter, Bruno 214
Wangemann, Adelbert Theodor Edward 96
Wangemann, Theo 120
Watts, Cecil 193
wax recording 59, 72, 80–83, 86, 87–88, 91–92, 149
Weekly Mail 141
Welch, Walter L. 162
Wellington, Duke of 38, 39
Wente, Edward Christopher 180
Wesley, John 34
Western Electric 179, 180, 182, 196
Western Union 48, 58, 64, 78
Westminster Abbey 179
Westrex 241
Wheatstone, Charles 42
Whiteman, Paul 158, 169–170, 177–178, 179–180, 185
Wile, Raymond R. 130
Wilhelm II, Kaiser 120
Wilkinson, Kenneth 203
Williams, Amy 164
Williams, Edmund Trevor Lloyd Wynne 136
Williams, Trevor 162
Willis, Percy 118–120
Winner records 175
Wireless Telegraph Company 193
Wonderful Talking Machine 37–39
World, The 76
Wright, John 247
Wurth, Charles 56–57

Yates, Edmund Hodgson 114, 115
Young, Thomas 24

zinc discs 129, 130, 131
Zonophone 134, 151, 182